Chapman & Hall/CRC Mathematical and Com

Introduction to Bioinformatics

CHAPMAN & HALL/CRC
Mathematical and Computational Biology Series

Aims and scope:

This series aims to capture new developments and summarize what is known over the whole spectrum of mathematical and computational biology and medicine. It seeks to encourage the integration of mathematical, statistical and computational methods into biology by publishing a broad range of textbooks, reference works and handbooks. The titles included in the series are meant to appeal to students, researchers and professionals in the mathematical, statistical and computational sciences, fundamental biology and bioengineering, as well as interdisciplinary researchers involved in the field. The inclusion of concrete examples and applications, and programming techniques and examples, is highly encouraged.

Series Editors

Alison M. Etheridge
Department of Statistics
University of Oxford

Louis J. Gross
Department of Ecology and Evolutionary Biology
University of Tennessee

Suzanne Lenhart
Department of Mathematics
University of Tennessee

Philip K. Maini
Mathematical Institute
University of Oxford

Shoba Ranganathan
Research Institute of Biotechnology
Macquarie University

Hershel M. Safer
Weizmann Institute of Science
Bioinformatics & Bio Computing

Eberhard O. Voit
The Wallace H. Couter Department of Biomedical Engineering
Georgia Tech and Emory University

Proposals for the series should be submitted to one of the series editors above or directly to:
CRC Press, Taylor & Francis Group
24-25 Blades Court
Deodar Road
London SW15 2NU
UK

Published Titles

Cancer Modelling and Simulation
Luigi Preziosi

Computational Biology: A Statistical Mechanics Perspective
Ralf Blossey

Computational Neuroscience: A Comprehensive Approach
Jianfeng Feng

Data Analysis Tools for DNA Microarrays
Sorin Draghici

Differential Equations and Mathematical Biology
D.S. Jones and B.D. Sleeman

Exactly Solvable Models of Biological Invasion
Sergei V. Petrovskii and Lian-Bai Li

Introduction to Bioinformatics
Anna Tramontano

An Introduction to Systems Biology: Design Principles of Biological Circuits
Uri Alon

Knowledge Discovery in Proteomics
Igor Jurisica and Dennis Wigle

Modeling and Simulation of Capsules and Biological Cells
C. Pozrikidis

Normal Mode Analysis: Theory and Applications to Biological and Chemical Systems
Qiang Cui and Ivet Bahar

Stochastic Modelling for Systems Biology
Darren J. Wilkinson

The Ten Most Wanted Solutions in Protein Bioinformatics
Anna Tramontano

Chapman & Hall/CRC Mathematical and Computational Biology Series

Introduction to Bioinformatics

Anna Tramontano

CRC Press
Taylor & Francis Group
Boca Raton London New York

CRC Press is an imprint of the
Taylor & Francis Group, an **informa** business

A CHAPMAN & HALL BOOK

Bioinformatica. Authorized translation from the Italian language edition published by Zanichelli.

First published 2007 by Chapman & hall/CRC Press

Published 2019 by CRC Press
Taylor & Francis Group
6000 Broken Sound Parkway NW, Suite 300
Boca Raton, FL 33487-2742

ISBN-13: 978-1-58488-569-6 (pbk)

Library of Congress Cataloging-in-Publication Data

Tramontano, Anna.
   Introduction to Bioinformatics / by Anna Tramontano.
      p. cm. -- (Mathematical and computational biology series)
   Includes bibliographical references and index.
   ISBN 1-58488-569-6
   1. Bioinformatics. I. Title.

QH324.2.T73 2006
572.80285--dc22   2006049140

Dedication

To my students and to my friend and mentor,
Maurizio Brunori, with unwavering respect and affection

Preface

Bioinformatics is a relatively recent discipline. The term first appeared in scientific papers at the beginning of the 1990s, but this fact can be misleading. Already in the 1960s, when research laboratories were able to afford computers with good graphical performance, the scientific community had started to use them to analyze biological data. From the point of view of informatics and biology, substantial progress has been made since then.

Hardware has become more efficient; the speed and graphical performance of personal computers now are astonishing compared with those of just 10 years ago. The process of developing software is made easier and more straightforward every day. Another important aspect has been the development and spread of the World Wide Web, an instrument that has changed our conception of communications and made a deep impact on the scientific community. Furthermore, biological data have been exponentially accumulating, thanks to new, powerful techniques available in every laboratory.

Bioinformatics has synergistically exploited new technologies, giving rise to a new scientific discipline, with its own history and even some revolutions. We can define bioinformatics as the science that uses the instruments of informatics to analyze biological data in order to formulate hypotheses about life. However, a scientific definition needs to be operational rather than semantic, and the aim of this book is to describe bioinformatics methods and tools in terms of their ability to help us solve biological problems.

Nevertheless, before starting we must address frequently asked questions: "Who is a bioinformatician?" "What is his or her cultural background?" "Should a bioinformatician be proficient in programming or should he or she know biology in detail?" The confusion arises from the fact that we use the word bioinformatician to indicate expert users of the available tools as well as developers of new and more powerful methods. Of course the background in the two cases could and should not be the same.

In the first case, it is important to have a good biological knowledge to be able to understand the results of bioinformatics analyses and, at the same time, to write simple programs. In the second, expertise in statistical methods, algorithms, and programming and a basic biological knowledge are required. In both cases, though, it is essential to understand biological problems and methods and the rational basis of available bioinformatics tools. This is a must if we wish to use them correctly or improve them.

Therefore, the aim of this book is to describe the rationale and the limitations of the methods and tools available to the biological community at large. It is directed to students who want to have an idea about what bioinformatics is before deciding whether it is worth getting deeper into the subject and to those who, having decided to pursue a career in experimental biology, want to have a grasp of the methods they will undoubtedly need during their research.

The questions that we will address concern ways of storing and (more importantly) retrieving the enormous amount of biological data produced every day (Chapter 1) and the methods to decrypt the information encoded by a genome (Chapter 2), to detect and exploit the evolutionary and functional relationships among biological elements (Chapters 3, 4, and 5), and to predict the three-dimensional structure of a protein (Chapters 6, 7, 8, and 9).

Chapter 10 offers a window to what the future holds, although in such a young and quickly evolving field as bioinformatics, we have learned that it is hard to predict what will come next, even in the very near future. This is proving to be even more difficult than predicting the structure and the function of a biological macromolecule!

The future will challenge us with new methodologies for tackling new and old problems, but some fundamental aspects will not change. We will always need to apply new methods to the same types of biological data, to implement them efficiently, and, most of all, to be aware of the power and limitations of these methods, in order to evaluate the meaningfulness of their results and extract information useful to solving biological problems.

Note: At the end of most chapters is a list of problems. Some of the input data for the problems can be downloaded in electronic format from the publisher's Web site, www.crcpress.com.

Author

Anna Tramontano studied physics at the University of Naples, Italy. She continued her research at the University of California San Francisco and became a staff scientist in the biocomputing programme of the European Molecular Biology Laboratory (EMBL) in Heidelberg. In 1990, Dr. Tramontano returned to Italy to work at the Merck Research Laboratories near Rome. In 2001, she returned to academia as Chair and Professor of Biochemistry at La Sapienza University, Rome where she continues today to pursue research in protein structure prediction and analysis.

Dr. Tramontano is a member of the European Molecular Biology Organization, the Scientific Council of Institute Pasteur-Fondazione Cenci Bolognetti, and serves on the organizing committee of the Critical Assessment of Techniques for Protein Structure Prediction (CASP) initiative. She is the director of two master's programs in bioinformatics, teaches at several universities, and is the coordinator in the European Permanent School in Bioinformatics.

Acknowledgments

While writing this book, I asked the advice of and help from many colleagues and friends. Special thanks go to Henriette Molinari, Arthur Lesk, Angelo Sironi, Armin Lahm, Sergio Ammendola, Renzo Bazzo, Valentina Cappello, Andrea Sbardellati, Domenico Cozzetto, and Adriana Miele, whose suggestions have been fundamental to the creation of the book. When there are errors, it is my fault; when there are none, it is thanks to them. Special thanks go to Domenico Raimondo and Alejandro Giorgetti, who preferred the challenge of the pixels and formats of several of the pictures in this book to the breathtaking beauty of the beaches on the island of Sardinia!

Table of Contents

1 The Data: Storage and Retrieval

GLOSSARY

cDNA: complementary DNA; a DNA molecule obtained by retrotranscribing an mRNA molecule into DNA

Constraints: restrictions of the possible values taken by a parameter, such as a distance, an angle, or a solid angle

Data bank/database: collection of information stored in a systematic way that can be accessed electronically and searched by various parameters

DNA (RNA) polymerase: enzyme responsible for catalyzing the synthesis of a new molecule of DNA (RNA) using a pre-existing template filament

Electrophoresis: a technique that allows the separation of charged compounds in an electric field

Entry: element of a database

EST: expressed sequence tags; DNA sequence obtained from the (partial) sequencing of a cDNA molecule

Hybridization: the process by which two complementary single-strand oligonucleotides associate

Nitrogenous bases: nitrogenous compounds (purines or pyrimidines) found in nucleosides, nucleotides, and nucleic acids

NMR: nuclear magnetic resonance; a technique that uses the interactions of nuclei with an external magnetic field to reconstruct their position in space, hence the structure of the molecule

Primer: RNA or DNA fragment complementary to a portion of the DNA region to be synthesized by the DNA polymerase

Ramachandran plot: plot that shows the theoretically allowed or experimentally observed combinations of ϕ and ψ angles in a polypeptide chain

Resolution, R factor, Rfree factor: parameters to evaluate the accuracy of the reconstruction of a macromolecular structure starting from x-ray diffraction data

Sequence pattern: a pattern of amino acids deemed to have a functional significance

SNP: single nucleotide polymorphism; naturally occurring variants that affect a single nucleotide (A, T, C, or G) in a genome

X-ray crystallography: technique that uses x-ray diffraction for the reconstruction of the three-dimensional positions of atoms inside molecular crystals

1.1 BASIC PRINCIPLES

Bioinformatics is a relatively young discipline that deals with the storage, retrieval, and analysis of biological data with informatics tools. Many branches of science use computers, databases, and algorithms, from weather forecasts to economics, from physics to linguistics. Each of them treats the data in different ways dictated by the nature of the data. From this perspective, we could define bioinformatics as the science that analyzes biological data with computer tools in order to formulate hypotheses on the processes underlying life.

Despite still being a more qualitative than quantitative science, modern biology has given bioinformatics a powerful push. Thanks to the development of new revolutionary experimental techniques, biological data have accumulated (and keep doing so) at an impressive pace. For example, we have available sequences of hundreds of genomes and data on the expression of thousands of genes in many cell types, and structural genomics projects are producing thousands of three-dimensional structures of proteins every year.

In first approximation, we can divide biological data into three main categories: sequence data, structural data, and functional data. The nature of the data and how their peculiar characteristics influence the organization of the databases where they are collected are discussed in this chapter.

A data bank should store data as they have originally been deposited so that they can be analyzed or reanalyzed with new or improved techniques at any time. These databases are usually called primary databases.

It is often useful to compute some properties of frequently used data and store them in different "derived" databases. This avoids the problem of repeating the same analysis, but it also implies that the derived database needs to be updated every time the primary data are updated. Ideally, this should happen in real time.

Biological databases contain different types of data, but, by and large, they refer to the same biological entities (genes or proteins). Therefore, it is essential to have instruments to connect and integrate the information contained in all biological databases and to allow the user to navigate easily from one to the other.

In this chapter, we will briefly discuss primary and derived databases as well as systems to integrate their contents. However, these are not static systems—they change to fit novel needs or include new types of data when they become available. Therefore, we will describe the main databases, which are not expected to change too much (at least in their basic principles). However, the reader should frequently consult the many available Internet resources that list databases and their developments, such as the home pages of the NCBI (National Center for Biological Information) and of the EBI (European Bioinformatics Institute).

For the same reason, the proposed solutions for the exercises at the end of this and other chapters can be used today to solve the problems, but they are not necessarily unique and we cannot guarantee that they remain the fastest route to the answer even in the near future.

1.2 THE DATA

1.2.1 THE POSTGENOMIC ERA

In 1953, Rosalind Franklin, Maurice Wilkins, James Watson, and Francis Crick published three papers in the scientific journal *Nature*, reporting the discovery of the structure of DNA. In 1962 Watson, Crick, and Wilkins became Nobel laureates. Rosalind Franklin was not among them because she died in 1958 at the age of 37. (The story is that she did not receive the prize because she had died; however, it is worth noting that none of the three Nobel laureates cited her fundamental contribution to the discovery in their Nobel speeches.)

By now it is common knowledge that DNA is a double helix made up of two strands of polynucleotides and that each nucleotide has one of four different *nitrogenous bases* (adenine, cytosine, guanine, and thymine). The structure revealed that pairs of bases were complementary (A with T and C with G) and thus the sequence of one of the two filaments is necessary and sufficient to know the sequence of the complementary strand (Figure 1.1).

The consequences of this discovery are astounding. It is probably one of the very few cases in which the knowledge of a biological structure directly provided novel and exciting functional information. It was immediately clear how each DNA molecule is capable of replicating itself and originating two absolutely identical strands; the last paragraph of the paper by Watson and Crick reads: "It has not escaped our notice that the specific pairing we have postulated immediately suggests a possible copying mechanism for the genetic material." Although we had to wait about 20 years for the technical details, the DNA structure also contained the route to the methodology for obtaining the sequence of its bases (Figure 1.2).

Given the four different nucleotides, a single strand of DNA, and a *primer* that is a fragment of DNA able to *hybridize* with a portion of a single-stranded DNA (i.e., with a sequence complementary to a portion of the region of interest), enzymes (*polymerases*) are able to elongate the primer and synthesize *de novo* the complete complementary DNA strand. If we add a nucleotide modified in such a way that it can be incorporated in the correct place but prevents further elongation of the strand, the synthesis of the complementary strand will stop randomly at sites where the polymerase will add the modified nucleotide in place of the normal one.

Therefore, the newly synthesized strands will be variable in length but will always end in positions where the template strand contains a base complementary to the modified nucleotide. If we carry out four reactions in parallel, each with a different modified nucleotide, we will end up with four different reaction mixtures. In each mixture will be fragments whose length will depend upon the position of the incorporated modified base. By separating the fragments according to their length via *electrophoresis*, we can deduce the exact position of all the bases and hence can "read" the DNA sequence (Figure 1.2).

It is certainly striking that we can "see" the detailed atomic structure of such a complex macromolecule just using a few chemicals, an enzyme, an electrophoretic apparatus, and our knowledge of the DNA structure.

FIGURE 1.1 The DNA double helix.

A conceptually easy experiment is not necessarily as easy to perform in practice, but, luckily, this is not the case here. DNA sequencing techniques are routinely used in every molecular biology laboratory. More importantly, the procedure can be automated; a few years after Gilbert and Sanger received the Nobel Prize for describing two independent methods for sequencing DNA, robots have been designed and produced to sequence the genome of humans and other animal and plant species automatically. As of today, sequencing of a bacterial genome of average size can take no longer than 1 or 2 weeks.

Here we are, living in the postgenomic era—a time when all of us, as scientists and as citizens of the world, must face the consequence of having learned how to interpret and manipulate a genome, a wonderful but also worrisome achievement.

Many are the medical and pharmacological implications of genome sequencing. In theory, we will be able to better understand the molecular bases of diseases and detect them early enough to devise more appropriate and efficacious therapies. We might also correlate genome features with the probability of being affected by certain diseases and this might lead us to modify our lifestyles in order to reduce the risks of their occurrence. We might catalogue patients on the basis of the probability that they react positively to a given medical treatment, thus reducing the problems of side effects or inefficacy of treatments, and the list could continue.

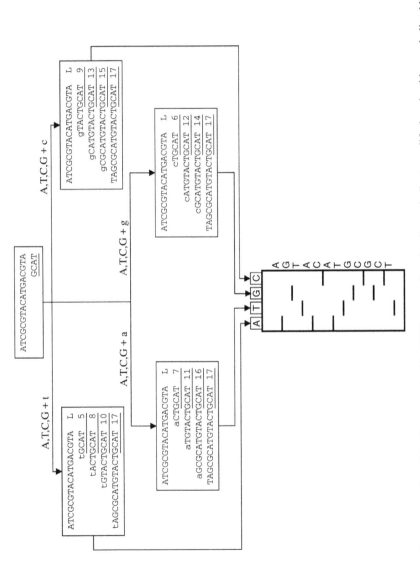

FIGURE 1.2 Schematic representation of the Sanger method for DNA sequencing. Small caps indicate modified nucleotides, underlined letters indicate the primer sequences.

However, we cannot ignore the risk that genetic information can be misused in unethical or dangerous ways. This book is aimed at unraveling the scientific and technical aspects of bioinformatics and will not deal with the ethical aspects of genomic research. It is nonetheless important to know that the risk of misuse does exist. The women and men living in the postgenomic era must face this new reality and carefully evaluate the advantages and problems connected with our discoveries as well as their impacts on society.

However, their impact on bioinformatics is obvious: We need to handle a huge amount of data and we must be able to store, classify, and retrieve them, discover their possible inter-relationships, and formulate new hypotheses. All of this must be done while new data are continuously produced in each laboratory of the world.

1.2.2 Nucleic Acid Data Banks

The striking simplicity of a DNA molecule is only apparent; in order to store the information that it contains, it is not sufficient to record its sequence and the organism to which it belongs. A database containing these raw data and not the annotations—that is, all the possible information known about the sequence—would barely be useful.

Nucleic acid databases indeed contain a plethora of experimental or deduced information on each of the sequences that they store. This implies that, if new properties of the biological entity encoded by a DNA sequence entry are unraveled, its annotations must be updated. This, at least in principle, needs to be done in agreement with the authors of the original deposition.

Now that so many annotated sequences are available, one has to face three important issues. First, the annotations related to a given sequence strongly depend on the annotations of other database entries. As will be discussed later in this book, similar amino acid sequences might code for proteins with similar function. If, for example, a previously deposited sequence is found to contain a previously nonidentified gene, its annotation must be modified, and this implies that all the related sequences and annotations must be changed accordingly. Unfortunately, this is not always possible.

It can happen that a new sequence is deposited that partially overlaps an existing database entry. Usually, both entries are stored; however, this might (and does) create problems for the users of the database. Moreover, the sequence of a given organism is not unique; *polymorphisms* and variants exist, and sometimes they are very frequent, as in viruses. A database must keep track and collect them all, although this redundancy does create confusion and problems for users. For example, if we search in a nucleotide database for sequences similar to hepatitis C, we will retrieve hundreds of sequences; none of them will directly suggest the functional properties of the viral proteins since they all belong to naturally occurring variants of the virus. Despite these known limitations, nucleotide databases are an invaluable tool in biology. The EMBL Sequence Data Bank, GenBank (Figure 1.3), and DDBJ are the most used.

Among the annotations of an entry for a nucleotide sequence encoding for a protein are the position of the start and end codons of the translation, the amino acid sequence of the product, and references to related entries in different databases. The translation of the gene into amino acids is the most frequent source of information for protein sequences.

It is well known that a genomic sequence contains coding regions for proteins and untranslated regions (such as introns, enhancers, intervening sequences, etc.). The messenger RNA (mRNA) sequence only contains the translated regions. It is possible to extract the molecules of mRNA from a population of cells, enzymatically retrotranscribe them into DNA molecules (*cDNA*), and sequence them. Several laboratories have started to sequence fragments of cDNA from different organisms and different cell types (even if with a higher percentage of error than that accepted for genomic sequences). These data are extremely useful in detecting coding regions within genomic DNA sequences, as we will see in the next chapter, and therefore are collected in a special database called *EST* (expressed sequence tags). An example of an *entry* is shown in Figure 1.4.

The position and types of polymorphisms are very important from a medical point of view. Specialized SNPs (single nucleotide polymorphism) databases collect data coming from different laboratories specializing in the study of (Figure 1.5).

The databases described previously are "gene" oriented; that is, they contain entries for each known gene. However, genome sequences provide a natural framework about which to organize biological data and a few systems—the most widely used are Ensembl and the Genome Browser at UCSD—exploit this fact, providing a very useful genome-oriented view of the data.

The aims of these systems are to provide an accurate, automatic analysis of genome data, releasing the data and the results of the analysis into the public domain immediately via the Web. An important part of these systems is an automatic gene identification system based upon several strategies that we will discuss in the next chapter, as well as the simultaneous visualization of genes, annotations, EST, SNP data, the location of disease-related mutations, and many more features in the context of a genomic location.

Unsurprisingly, the availability of the genomes of several species has led to the development of methods to compare them and navigate from one to the other. Genomics databases provide such capabilities; for example, they allow users to see the relationship between the chromosomal location of homologous genes in different species.

1.2.3 PROTEIN SEQUENCE DATA BANKS

There are experimental methods to sequence proteins directly, but they are not as easy as those used for sequencing nucleic acids. Indeed, the large majority of available protein sequences are derived from the translation of the genes encoding

```
LOCUS        BTTRYPAP        805 bp     mRNA           MAM        06-JUN-1991
DEFINITION   Bovine mRNA for pancreatic anionic trypsinogen.
ACCESSION    X54703
VERSION      X54703.1   GI:829
KEYWORDS     trypsin.
SOURCE       cow.
  ORGANISM   Bos taurus
             Eukaryota; Metazoa; Chordata; Craniata; Vertebrata; Euteleostomi;
             Mammalia; Eutheria; Cetartiodactyla; Ruminantia; Pecora; Bovoidea;
             Bovidae; Bovinae; Bos.
REFERENCE    1  (bases 1 to 805)
  AUTHORS    Le Huerou,I.
  TITLE      Direct Submission
  JOURNAL    Submitted (28-AUG-1990) Le Huerou I., Centre National de la ...
             Recherche Scientifique, Centre de Biochimie et de Biologie
             Moleculaire, CBM3, 31, Chemin Joseph Aiguier, B.P. 71, 13402
             Marseille Cedex 9, France
REFERENCE    2  (bases 1 to 805)
  AUTHORS    Le Huerou,I., Wicker,C., Guilloteau,P., Toullec,R. and
             Puigserver,A.
  TITLE      Isolation and nucleotide sequence of cDNA clone for bovine
             pancreatic anionic trypsinogen. Structural identity within the
             trypsin family
  JOURNAL    Eur. J. Biochem. 193 (3), 767-773 (1990)
  MEDLINE    91065383
COMMENT      Data kindly reviewed (05-FEB-1991) by Huerou I.
FEATURES             Location/Qualifiers
     source          1..805
                     /organism="Bos taurus"
                     /strain="Friesian Holstein"
                     /db_xref="taxon:9913"
                     /dev_stage="adult"
                     /tissue_type="pancreas"
                     /clone_lib="pUC9"
                     /clone="PB10"
     mRNA            <1..805
     CDS             9..752
                     /EC_number="3.4.21.4"
                     /codon_start=1
                     /product="trypsinogen anionic precursor"
                     /protein_id="CAA38513.1"
                     /db_xref="GI:830"
                     /db_xref="SWISS-PROT:Q29463"
                     /translation="MHPLLILAFVGAAVAFPSDDDDKIVGGYTCAENSVPYQVSLNAG
                     YHFCGGSLINDQWVVSAAHCYQYHIQVRLGEYNIDVLEGGEQFIDASKIIRHPKYSSW
                     TLDNDILLIKLSTPAVINARVSTLLLPSACASAGTECLISGWGNTLSSGVNYPDLLQC
                     LVAPLLSHADCEASYPGQITNNMICAGFLEGGKDSCQGDSGGPVACNGQLQGIVSWGY
                     GCAQKGKPGVYTKVCNYVDWIQETIAANS"
     sig_peptide     9..53
     misc_feature    54..77
                     /product="activation peptide"
     mat_peptide     78..749
                     /EC_number="3.4.21.4"
                     /product="trypsin"
     polyA_signal    783..788
     polyA_site      805
BASE COUNT       157 a     255 c      220 g     173 t
ORIGIN
        1 tctccaccat gcatccctg cttatccttg cctttgtggg agctgctgtg gctttcccct
       61 cggacgacga tgacaagatc gtcgggggct acacctgcgc agagaattcc gtcccttacc
      121 aggtgtccct gaatgctggc taccacttct gcgggggctc cctcatcaat gaccagtggg
      181 tggtgtccgc ggctcactgc taccagtacc acatccaggt gaggctggga gaatacaaca
      241 ttgatgtctt ggagggtggt gagcagttca tcgatgcgtc caagatcatc cgccacccca
      301 agtacagcag ctggactctg gacaatgaca tcctgctgat caaactctcc acgcctgcgg
      361 tcatcaatgc ccgggtgtcc accttgctgc tgcccagtgc ctgtgcttcc gcaggcacag
      421 agtgcctcat ctccggctgg ggcaacaccc tgagcagtgg cgtcaactac ccggacctgc
      481 tgcaatgcct ggtggcccccg ctgctgagcc acgccgactg tgaagcctca taccctggac
      541 agatcactaa caacatgatc tgcgctggct tcctggaagg aggcaaggat tcctgccagg
      601 gtgactctgg cggccctgtg gcttgcaacg gacagctcca gggcattgtg tcctgggggct
      661 acggctgtgc ccagaagggc aagcctgggg tctacaccaa ggtctgcaac tacgtggact
      721 ggattcagga gaccatcgcc gccaacagct gaagccctgc ccctctctgc catcattatg
      781 ctaataaaga gactgctctt cctgc
//
```

FIGURE 1.3 Example of an entry of the GENBANK nucleotide database. Note the "FEA-TURES" fields, which contain the classification of the enzyme (EC number 3.4.21.4), the name of the gene product, references to related entries in other databases, the amino acid translation, and the position of the different elements of the gene.

them. An entry in a protein database is organized in a similar way as those in nucleotide databases: It includes functional annotation and the protein sequence in one-letter-code. The chemical structure and the names of the 20 amino acids

```
LOCUS       AI589465              416 bp    mRNA    linear   EST 14-MAY-1999
DEFINITION  tr76d10.x1 NCI_CGAP_Pan1 Homo sapiens cDNA clone IMAGE:2224243 3',
            mRNA sequence.
ACCESSION   AI589465
VERSION     AI589465.1  GI:4598513
KEYWORDS    EST.
SOURCE      Homo sapiens (human)
  ORGANISM  Homo sapiens
            Eukaryota; Metazoa; Chordata; Craniata; Vertebrata; Euteleostomi;
            Mammalia; Eutheria; Euarchontoglires; Primates; Haplorrhini;
            Catarrhini; Hominidae; Homo.
REFERENCE   1  (bases 1 to 416)
  AUTHORS   NCI-CGAP http://www.ncbi.nlm.nih.gov/ncicgap.
  TITLE     National Cancer Institute, Cancer Genome Anatomy Project (CGAP),
            Tumor Gene Index
  JOURNAL   Unpublished (1997)
COMMENT     Contact: Robert Strausberg, Ph.D.
            Email: cgapbs-r@mail.nih.gov
            Life Technologies catalog #: 11548-013
             DNA Sequencing by: Washington University Genome Sequencing Center
             Clone distribution: NCI-CGAP clone distribution information can be
            found through the I.M.A.G.E. Consortium/LLNL at:
            www-bio.llnl.gov/bbrp/image/image.html
            Insert Length: 4111    Std Error: 0.00
            Seq primer: -40UP from Gibco
            High quality sequence stop: 372
            POLYA=No.
FEATURES             Location/Qualifiers
     source          1..416
                     /organism="Homo sapiens"
                     /mol_type="mRNA"
                     /db_xref="taxon:9606"
                     /clone="IMAGE:2224243"
                     /tissue_type="adenocarcinoma"
                     /lab_host="DH10B"
                     /clone_lib="NCI_CGAP_Pan1"
                     /note="Organ: pancreas; Vector: pCMV-SPORT6; Site_1: SalI;
                     Site_2: NotI; Cloned unidirectionally.  Primer: Oligo dT.
                     Average insert size 1.72 kb. Life Technologies catalog #:
                     11548-013"
ORIGIN
        1 ttttcagagc aaatgtttta ttaacagatt tttctccagt agtcttgaaa atctcccctg
       61 ggggagaaaa ctggcaaaag ggggcctgag ggttgttgga agaaatgaaa agtacatagg
      121 atgctatttt gaagaactga tttgtcagaa gccactttat tttaacaata atgtgactgc
      181 acctggcaca tgagacaggc tctgggcctg tttctcaatg ttgtgaaaag caggacagtc
      241 agttcccatg gagtagatta cagcgtgttg gacaaggaca ttccnccgcg tgaaggcaca
      301 ctccatggag ccagtggggc caccatntca cgtgtgcaag ctgtggcctt ggagagaaca
      361 tgggcagggc ttccccagcc tctactatac cccagcttta ctccatgatc acaaca
```

FIGURE 1.4 Example of an EST database entry. Information on the experimental clone (which can be requested) is listed together with information on the sequence.

in full and their (extremely useful) three-letter and one-letter codes are shown in Table 1.1.

As we mentioned, annotations represent a very important part of the databases; therefore, the "best" database is, by definition, the one with the most accurate, updated, and organized annotations, as is the case of the SwissProt database (Figure 1.6[abc]). In a SwissProt entry, there are references to other databases: primary, such

rs1140781 [Homo sapiens]
>gnl|dbSNP|rs1140781
rs=1140781|pos=51|len=101|taxid=9606|mol="cDNA"|class=1|alleles="A/C"|build=86
TGTCTCTATT CAGGAAGAGA TAGTAGGTGA TTTCAAATCG GAGAAGTCCA
[A/C]
CGGGGAGCTA AGTGAATCTC CTGGAGCTGG AAAAGGAGCA TCTGGCTCAA

FIGURE 1.5 Example of an SNP entry. The polymorphism is indicated by the [A/C] symbol.

TABLE 1.1
Structure and Classification of the 20 Amino Acids Used in Nature

One Letter	Three Letters	Name	Linear Structure
A	Ala	Alanine	CH3-CH(NH2)-COOH
C	Cys	Cysteine	HS-CH2-CH(NH2)-COOH
D	Asp	Aspartic acid	HOOC-CH2-CH(NH2)-COOH
E	Glu	Glutamic acid	HOOC-(CH2)2-CH(NH2)-COOH
F	Phe	Phenylalanine	Ph-CH2-CH(NH2)-COOH
G	Gly	Glycine	NH2-CH2-COOH
H	His	Histidine	NH-CH=N-CH=C-CH2-CH(NH2)-COOH
I	Ile	Isoleucine	CH3-CH2-CH(CH3)-CH(NH2)-COOH
K	Lys	Lysine	H2N-(CH2)4-CH(NH2)-COOH
L	Leu	Leucine	(CH3)2-CH-CH2-CH(NH2)-COOH
M	Met	Methionine	CH3-S-(CH2)2-CH(NH2)-COOH
N	Asn	Asparagine	H2N-CO-CH2-CH(NH2)-COOH
P	Pro	Proline	NH-(CH2)3-CH-COOH
Q	Gln	Glutamine	H2N-CO-(CH2)2-CH(NH2)-COOH
R	Arg	Arginine	HN=C(NH2)-NH-(CH2)3-CH(NH2)-COOH
S	Ser	Serine	HO-CH2-CH(NH2)-COOH
T	Thr	Threonine	CH3-CH(OH)-CH(NH2)-COOH
V	Val	Valine	(CH3)2-CH-CH(NH2)-COOH
W	Trp	Tryptophan	Ph-NH-CH=C-CH2-CH(NH2)-COOH
Y	Tyr	Tyrosine	HO-Ph-CH2-CH(NH2)-COOH

as the EMBL, and derived, such as PRINTS and PROSITE. Moreover, the description
of the protein (FT lines) includes details about its active site, the presence and
location of signal peptides and disulfide bridges, etc.

Experts manually verify the annotations in the SwissProt database, and this is clearly
a great advantage. On the other hand, given the speed at which data accumulate, careful

```
ID   RNBR_BACAM      STANDARD;       PRT;    157 AA.
AC   P00648;
DT   21-JUL-1986, integrated into UniProtKB/Swiss-Prot.
DT   01-JUL-1989, sequence version 2.
DT   21-FEB-2006, entry version 64.
DE   Ribonuclease precursor (EC 3.1.27.-) (Barnase) (RNase Ba).
OS   Bacillus amyloliquefaciens.
OC   Bacteria; Firmicutes; Bacillales; Bacillaceae; Bacillus.
OX   NCBI_TaxID=1390;
RN   [1]
RP   NUCLEOTIDE SEQUENCE [GENOMIC DNA].
RX   MEDLINE=86165864; PubMed=3007290; DOI=10.1016/0378-1119(85)90045-9;
RA   Paddon C.J., Hartley R.W.;
RT   "Cloning, sequencing and transcription of an inactivated copy of
RT   Bacillus amyloliquefaciens extracellular ribonuclease (barnase).";
RL   Gene 40:231-239(1985).
RN   [2]
RP   NUCLEOTIDE SEQUENCE [GENOMIC DNA] OF 48-157.
RX   MEDLINE=89012012; PubMed=3050134;
RA   Hartley R.W.;
RT   "Barnase and barstar. Expression of its cloned inhibitor permits
RT   expression of a cloned ribonuclease.";
RL   J. Mol. Biol. 202:913-915(1988).
RN   [3]
RP   PROTEIN SEQUENCE OF 48-157.
RX   MEDLINE=72164019; PubMed=4553460;
RA   Hartley R.W., Barker E.A.;
RT   "Amino-acid sequence of extracellular ribonuclease (barnase) of
RT   Bacillus amyloliquefaciens.";
RL   Nature New Biol. 235:15-16(1972).
RN   [4]
RP   STRUCTURE BY NMR.
RX   MEDLINE=91363360; PubMed=1888730;
RA   Bycroft M., Ludvigsen S., Fersht A.R., Poulsen F.M.;
RT   "Determination of the three-dimensional solution structure of barnase
RT   using nuclear magnetic resonance spectroscopy.";
RL   Biochemistry 30:8697-8701(1991).
RN   [5]
RP   STRUCTURE BY NMR OF 48-83.
RX   MEDLINE=92235829; PubMed=1569554;
RA   Sancho J., Neira J.L., Fersht A.R.;
RT   "An N-terminal fragment of barnase has residual helical structure
RT   similar to that in a refolding intermediate.";
RL   J. Mol. Biol. 224:749-758(1992).
RN   [6]
RP   STRUCTURE BY NMR OF 48-157.
RX   MEDLINE=98372644; PubMed=9708913; DOI=10.1016/S0014-5793(98)00765-0;
RA   Reibarkh M.Y.A., Nolde D.E., Vasilieva L.I., Bocharov E.V.,
RA   Shulga A.A., Kirpichnikov M.P., Arseniev A.S.;
RT   "Three-dimensional structure of binase in solution.";
RL   FEBS Lett. 431:250-254(1998).
RN   [7]
RP   X-RAY CRYSTALLOGRAPHY (1.9 ANGSTROMS).
RX   MEDLINE=91218173; PubMed=2023257;
RA   Baudet S., Janin J.;
RT   "Crystal structure of a barnase-d(GpC) complex at 1.9-A resolution.";
RL   J. Mol. Biol. 219:123-132(1991).
RN   [8]
RP   X-RAY CRYSTALLOGRAPHY (2.6 ANGSTROMS) OF COMPLEX WITH BARNASE.
RA   Guillet V., Lapthorn A., Hartley R.W., Mauguen Y.;
RT   "Recognition between a bacterial ribonuclease, barnase, and its
RT   natural inhibitor, barstar.";
RL   Structure 1:165-177(1993).
```

(a)

FIGURE 1.6 Example of a SwissProt entry.

```
RN   [9]
RP   X-RAY CRYSTALLOGRAPHY (1.7 ANGSTROMS).
RX   MEDLINE=21911532; PubMed=11914482; DOI=10.1107/S0907444902001567;
RA   Vaughan C.K., Harryson P., Buckle A.M., Fersht A.R.;
RT   "A structural double-mutant cycle: estimating the strength of a buried
RT   salt bridge in barnase.";
RL   Acta Crystallogr. D 58:591-600(2002).
RN   [10]
RP   REVIEW.
RX   MEDLINE=90162921; PubMed=2696173; DOI=10.1016/0968-0004(89)90104-7;
RA   Hartley R.W.;
RT   "Barnase and barstar: two small proteins to fold and fit together.";
RL   Trends Biochem. Sci. 14:450-454(1989).
CC   -!- FUNCTION: Hydrolyzes phosphodiester bonds in RNA, poly- and
CC       oligoribonucleotides resulting in 3'-nucleoside monophosphates via
CC       2',3'-cyclophosphate intermediates.
CC   -!- SUBCELLULAR LOCATION: Secreted protein.
CC   -!- BIOTECHNOLOGY: Introduced by genetic manipulation and expressed in
CC       male sterile maize and rape by Plant Genetic Systems and in
CC       radicchio rosso by Bejo Zaden. Barnase expressed in transgenic
CC       plants will specifically destroy the tissue(s) where it is
CC       expressed. After cell transformation, protein expression and plant
CC       regeneration, the active nuclease destroys the pollen producing
CC       organs thus rendering the plant sterile.
CC   -!- SIMILARITY: Belongs to the ribonuclease N1/T1 family.
CC   -----------------------------------------------------------------------
CC   Copyrighted by the UniProt Consortium, see http://www.uniprot.org/terms
CC   Distributed under the Creative Commons Attribution-NoDerivs License
CC   -----------------------------------------------------------------------
DR   EMBL; M14442; AAA86441.1; -; Genomic_DNA.
DR   EMBL; X12871; CAA31365.1; -; Genomic_DNA.
DR   PIR; A24038; NRBSN.
DR   PDB; 1A2P; X-ray; A/B/C=48-157.
DR   PDB; 1B20; X-ray; A/B/C=48-157.
DR   PDB; 1B21; X-ray; A/B/C=48-157.
DR   PDB; 1B27; X-ray; A/B/C=48-157.
DR   PDB; 1B2S; X-ray; A/B/C=48-157.
DR   PDB; 1B2U; X-ray; A/B/C=48-157.
DR   PDB; 1B2X; X-ray; A/B/C=48-157.
DR   PDB; 1B2Z; X-ray; A/B/C=48-157.
DR   PDB; 1B3S; X-ray; A/B/C=48-157.
DR   PDB; 1BAN; X-ray; A/B/C=48-157.
DR   PDB; 1BAO; X-ray; A/B/C=48-157.
DR   PDB; 1BGS; X-ray; A/B/C=48-157.
DR   PDB; 1BNE; X-ray; A/B/C=48-157.
DR   PDB; 1BNF; X-ray; A/B/C=48-157.
DR   PDB; 1BNG; X-ray; A/B/C=48-157.
DR   PDB; 1BNI; X-ray; A/B/C=48-157.
DR   PDB; 1BNJ; X-ray; A/B/C=48-157.
DR   PDB; 1BNR; NMR; @=48-157.
DR   PDB; 1BNS; X-ray; A/B/C=48-157.
DR   PDB; 1BRG; X-ray; A/B/C=50-157.
DR   PDB; 1BRH; X-ray; A/B/C=48-157.
DR   PDB; 1BRI; X-ray; A/B/C=48-157.
DR   PDB; 1BRJ; X-ray; A/B/C=48-157.
DR   PDB; 1BRK; X-ray; A/B/C=48-157.
DR   PDB; 1BRN; X-ray; L/M=48-157.
DR   PDB; 1BRS; X-ray; A/B/C=48-157.
DR   PDB; 1BSA; X-ray; A/B/C=48-157.
DR   PDB; 1BSB; X-ray; A/B/C=48-157.
DR   PDB; 1BSC; X-ray; A/B/C=48-157.
DR   PDB; 1BSD; X-ray; A/B/C=48-157.
DR   PDB; 1BSE; X-ray; A/B/C=48-157.
```

(b)

FIGURE 1.6 (Continued.)

```
DR   PDB; 1FW7; NMR; A=48-157.
DR   PDB; 1RNB; X-ray; A=48-157.
DR   PDB; 1YVS; X-ray; @=48-157.
DR   PDB; 2C4B; X-ray; A/B=-.
DR   LinkHub; P00648; -.
DR   InterPro; IPR001887; Barnase.
DR   InterPro; IPR000026; Ribonuc_N1T1.
DR   Pfam; PF00545; Ribonuclease; 1.
DR   PRINTS; PR00117; BARNASE.
KW   3D-structure; Direct protein sequencing; Endonuclease;
KW   Genetically modified food; Hydrolase; Nuclease; Signal.
FT   SIGNAL          1      34
FT   PROPEP         35      47
FT                                     /FTId=PRO_0000030828.
FT   CHAIN          48     157          Ribonuclease.
FT                                     /FTId=PRO_0000030829.
FT   ACT_SITE      120     120          Proton acceptor.
FT   ACT_SITE      149     149          Proton donor.
FT   MUTAGEN       149     149          H->Q: Loss of activity.
FT   STRAND         53      53
FT   HELIX          54      64
FT   STRAND         65      66
FT   TURN           69      70
FT   STRAND         71      72
FT   HELIX          74      80
FT   TURN           81      81
FT   HELIX          84      86
FT   TURN           87      87
FT   HELIX          89      92
FT   STRAND         93      93
FT   TURN           94      95
FT   STRAND         97     103
FT   TURN          106     107
FT   STRAND        110     110
FT   TURN          114     115
FT   STRAND        118     122
FT   STRAND        124     124
FT   STRAND        127     129
FT   STRAND        132     132
FT   STRAND        134     138
FT   TURN          139     140
FT   STRAND        143     148
FT   TURN          149     150
FT   STRAND        151     151
FT   STRAND        154     155
SQ   SEQUENCE      157 AA;  17473 MW;  42B6669F9B0D7FBA CRC64;
     MMKMEGIALK KRLSWISVCL LVLVSAAGML FSTAAKTETS SHKAHTEAQV INTFDGVADY
     LQTYHKLPDN YITKSEAQAL GWVASKGNLA DVAPGKSIGG DIFSNREGKL PGKSGRTWRE
     ADINYTSGFR NSDRILYSSD WLIYKTTDHY QTFTKIR
```

(c)

FIGURE 1.6 (Continued.)

manual annotation of every entry is not a feasible route. Most of the annotations in other protein databases are derived using automatic methods, as we will discuss.

Other protein sequence databases are TrEMBL and PIR, but there is no need to learn their location or format because they have been recently integrated into a resource called the Universal Protein Resource (UniProt). UniProt comprises three components. The **UniProt Knowledgebase (UniProtKB)** is the central access point for extensive curated protein information, including function, classification, and cross-reference. The **UniProt Reference Clusters (UniRef)** databases combine closely related sequences into a single record to speed searches. The **UniProt Archive (UniParc)** is a comprehensive repository reflecting the history of all protein sequences. The system includes data from Swiss-Prot, TrEMBL, and PIR; the Uni-Prot/Swiss-Prot entries, derived from the SwissProt database, are manually curated entries.

1.2.4 PROTEIN STRUCTURE DATABASES

Two experimental techniques can be used to determine the three-dimensional structure of macromolecules at an atomic level: *x-ray crystallography* and *nuclear magnetic resonance (NMR)*. These techniques are complex from theoretical and experimental points of view, and their description is out of the scope of this book. However, it is impossible to use the data they produce correctly without learning some very basic facts about the techniques.

X-ray diffraction is based on the observation that an ordered ensemble of molecules, arranged in a crystal lattice, can diffract x-rays when hit by an incident beam. As Bragg said in his Nobel Prize acceptance speech in 1922:

> The rays are diffracted by the electrons grouped around the centre of each atom. In some directions the atoms conspire to give a strong scattered beam, in others their effects almost annul each other by interference. The exact arrangement of the atoms is to be deduced by comparing the strength of the refle[ct]ions from different faces and in different orders.

In other words, the electromagnetic waves (x-rays) are dispersed by the electrons in the molecule and interfere with each other, giving rise to a pattern of maxima and minima of intensities. The pattern depends on the amplitude and phase of the interacting waves, which, in turn, depend on the position of the electrons (and hence the atoms) in the ordered molecules of the crystal. The electrons of each molecule, whose position depends upon the three-dimensional structure of the molecule, diffract the incident beam. All the molecules in an ordered crystal reinforce this wave, acting as an amplifier, and produce a diffraction pattern that can be measured experimentally.

Unfortunately, photographic films and electronic detectors can only be used to measure the intensities, but not the phases, of the diffracted waves. Therefore, in order to reconstruct the image that generated a given diffraction pattern, the phases of the diffracted waves need to be computed—a problem that can be solved with different experimental and computational techniques that we will not discuss.

In a typical x-ray crystallography experiment we need to:

- grow a crystal of the target molecule
- irradiate it with a source of x-rays
- measure the intensity of the diffracted waves (from which we can derive their amplitude)
- compute the phases of the diffracted waves
- combine the phases with the amplitudes in order to calculate the electron density of the molecule
- model the structure of the molecule in such a way that its atoms fit the computed electron density

The final coordinates of the protein (or nucleic acid) are deposited in a database called Protein Data Bank (PDB), which will be described in more detail in Chapter 7. Here we will only briefly describe some parameters, reported in the database entry describing a protein structure, that are directly related to the quality of the experimental data.

The *resolution*, expressed in angstroms (Å), is a measure of the ratio between the number of calculated parameters (the atom positions) and the amount of measured data (the diffracted beam amplitudes). The lower its value is, the better is the quality of the structure. A resolution of about 3 Å allows secondary structure elements and the direction of the polypeptide chain to be distinguished; a resolution of 2.5 Å allows the side chain to be built with reasonable precision. At around 1.0 Å, we can even see the electron density of the electron-poor hydrogen atoms.

Once the model of the macromolecule is completely built, we can back-compute its expected electron density map. The *R factor* is an indication of how much the theoretical map differs from the experimentally observed one. This factor is linked to the resolution. A 3.0-Å structure of good quality is expected to have an R factor lower than 30%; for a 2.0-Å resolution structure, the factor should be lower than 20%, and so on. Another quality-control parameter is the *Rfree*, which indicates the agreement between the computed electron density and a set of experimental data that were left aside and not used in the calculation of the structure. It is somehow more objective than the R factor. For a correctly calculated structure, the ratio R/Rfree should be higher than 80%.

Nuclear magnetic resonance is another very useful technique to determine the structure of macromolecules. The basic principle comes from the observation that several nuclei (e.g., H, ^{13}C, ^{15}N) have an intrinsic magnetic moment. If we place a concentrated homogeneous solution of a protein (or of a nucleic acid) in a very high magnetic field, the nuclei of its atoms will assume one of their allowed spin orientations, the energy of which depends on the external field and the chemical environment. By applying a magnetic field in the radio frequencies to the sample, we can measure a strong energy absorbance when the frequency of the external field exactly matches the energy difference between two allowed spin orientations. Each amino acid will absorb at a set of frequencies (spin system) that depends upon its chemical structure. Given the amino acid sequence of the protein under study, we can identify the signals produced by each of its amino acids. If two atoms are close in space, we can also measure magnetic interactions between their spins. The intensity of the interaction rapidly decays with the reciprocal of the sixth power of the distance.

A set of NMR experiments allows the identification of pairs of atoms (for example, hydrogens) that are close to each other (less than 6 Å) and therefore influence each other. Subsequently, a number of possible protein structures with the correct stereochemistry that are compatible with the measured distance ranges can be generated. In general, the generated conformations are between 10 and 50, depending upon the number of measured distance *constraints*, which in turn depends upon the rigidity of the protein in solution. If a given region is very mobile, the nuclei of its atoms will not spend sufficient time next to other parts of the molecule to be influenced by them. In these cases, we cannot measure the interactions, but we gain invaluable information on the intrinsic mobility of the molecule.

NMR structures are deposited in the PDB as well. For each entry, a variable number of structures will be compatible with the data; the more similar they are the better is the structure defined. This is expressed by parameters, the root mean square deviation among them (r.m.s.d.), between the structures. We will discuss the r.m.s.d. in more detail in Chapter 7; here it suffices to say that it represents the average distance between equivalent atoms in the different structures. In some cases, only one "average" structure is present in the database.

Just a few comments are necessary about the advantages and shortcomings of the two methods—the subjects of a never ending debate. Is the structure of a protein in a crystal the same as in more physiological conditions? Generally, the answer is yes; structures solved by x-ray crystallography and nuclear magnetic resonance are essentially the same, within the limits of the experimental error. Protein crystals are full of solvent (and for this reason very fragile) and it is usually possible to follow enzymatic reactions in a crystal. NMR is limited by the dimension of the protein (even though the upper limit of the size of the molecules that can be studied keeps increasing every year) and gives a family of structures. This latter observation is more an advantage than a limitation, since it provides information about the dynamical behavior of the protein.

Both techniques can extract complementary information about biological macromolecules and thus represent extremely useful tools in modern biology.

1.2.5 PROTEIN INTERACTION DATABASES

Proteins do not act alone. Most cellular functions are mediated by transient or stable protein complexes, and their identification is of clear importance if we want to know the parts list of a biological system. Recent developments in technology, such as microarrays; yeast two-hybrid, high-throughput immunopurification experiments; and mass spectrometry, are generating a large amount of data on protein interactions. This, of course, calls for the development of databases to store and organize the information.

Protein interaction databases ideally should accommodate all types of protein interactions observed in biology—that is, metabolic and signaling pathways; morphogenic pathways involving proteins that participate in the same cellular function during a developmental process; and structural complexes and molecular machines in which many macromolecules are brought together.

Several biological databases are dedicated to protein interactions, usually with integrated tools for browsing. For example, DIP includes experimentally determined protein–protein interactions; BIND includes protein–protein, protein–RNA, protein–DNA, and protein–small molecule interactions; and MIPS contains compiled protein interaction data for yeast, etc. The diverse nature of protein interactions makes it impossible for interaction databases to have simple data structure and representation. On the other hand, they serve many important purposes: They are centralized data repositories that allow validation of protein interactions by comparing results, navigation of the protein-interaction network, discovery of new pathways and modes of regulation; and study of general properties of biological networks.

1.2.6 Derived Data Banks

All the databases described so far contain experimental data and annotations and are called "primary." Their content can be analyzed in many different ways, as described later. Nonetheless, some analyses are very useful and of general interest, and specialized research groups can perform them. Their results can be stored in databases and made available to the rest of the scientific community. These databases are called "derived."

In Table 1.2 some of the most frequently used derived databases are listed. Among them is the EST database that we discussed before. Some other databases, such as Pfam, Prints, and BLOCKS, group and align protein sequences (or their conserved regions in the case of BLOCKS) of evolutionarily related proteins. Prosite maintains and annotates all patterns found in sequences of proteins that have been correlated with a specific function. Protein structures are used to calculate secondary

TABLE 1.2
Some of the Most Frequently Used Derived Databases

Name	Derives From	Contains
EMEST	EST database	Groups and alignments of EST sequences
DSSP	PDB	Secondary structure assignments
HSSP	PDB and protein sequences databases	Alignment of proteins with known structure with all the similar sequences
FSSP, SCOP, CATH	PDB	Structural classification
3Dee	PDB	Protein domains definitions
Pfam, Prints, BLOCKS	Protein sequence databases	Alignments of proteins belonging to a homologous family
Prodom	Protein sequence databases	Alignment of protein domains with similar structure
Prosite	Protein sequence databases	Patterns
OMIM	Genomic databases	Genes and genetic diseases associated
LocusLink	Genomic databases	Genetic loci

structure elements (α-helices and β-sheets) and to generate DSSP (dictionary of secondary structures of proteins) or to group them according to their structural similarities (as in FSSP, SCOP, and CATH). For each protein of known structure, HSSP contains a list of proteins deemed to belong to the same evolutionary family together with an inferred evolutionary correspondence between their amino acids— that is, a sequence alignment.

1.2.7 INTEGRATION OF DATABASES

From a user point of view, an important problem deriving from the ever increasing number of databases is their integration. If we find the sequence of a gene of interest in a nucleic acid database, we would like to "jump" quickly to the corresponding entry in a protein sequence database; visualize its structure, if known; verify whether it belongs to a known family and whether it presents functional or structural peculiarities; and, finally, access the scientific publications describing its properties. Some moderately sophisticated systems (SRS and ENTREZ, for example) have been designed to this end and do allow navigating from one database to another, as shown schematically in Figure 1.7.

1.3 DATA QUALITY

Are there errors in databases? Yes, any database is bound to contain errors in the data or in the annotations. The former can be inherited from the original source (e.g., the experimental laboratory) or due to errors in data entry by the curators of the database (although nowadays most of the data are submitted directly by the authors in electronic form). Annotation errors may be due to the methodology used to annotate sequences, to error propagation from erroneous annotations of pre-existing entries, or to an incomplete or erroneous knowledge about the properties of the molecule at the time of deposition.

For example, we might have a two-domain protein. Each of the domains can have a different function, but only one might be known when the data are deposited. The known function might be assigned to proteins sharing a high degree of similarity with the original entry, even if the similarity is limited to the domain of unknown function (Figure 1.8).

In general, these problems could be solved by looking at the literature or by analyzing the data very carefully. However, this can be difficult if a large amount of data (such as the whole genome of a high organism) needs to be annotated. In this latter case, we do need to use automatic methods.

1.4 DATA REPRESENTATION

DNA and RNA sequences can be simply represented as the sequence of their bases, indicated by one of the five letters A (which stands for adenine), C (cytosine), T (thymine), G (guanosine), and U (uracil). Amino acids can be represented by their three-letter or one-letter codes (for obvious reasons, the latter are used in databases) (Table 1.1).

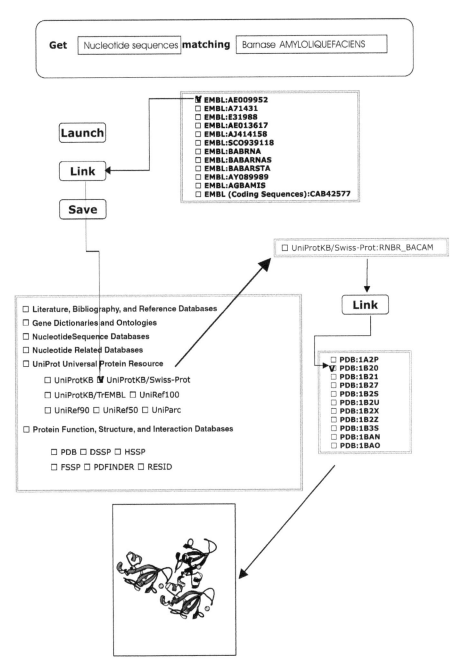

FIGURE 1.7 Example of the use of an integrated system for accessing biological databases. In this example, the SRS system is used to search the EMBL nucleic acids sequence database for all entries containing the word "barnase." Among all retrieved sequences, one is selected and used to query SwissProt and PDB for all related entries. The system also allows the structure of the protein, if known, to be displayed and manipulated interactively (rotated, translated, and resized).

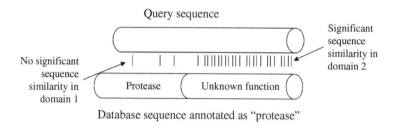

FIGURE 1.8 Example of a possible source of errors in the annotation of a database entry.

Often, conserved elements in a protein sequence have a precise biological role. For example, sequence and structure analyses revealed that most of the ATP- or GTP-binding proteins take advantage of a flexible loop between a β-strand and an α-helix to recognize the nucleotide. Generally, the loop starts with a small amino acid (alanine or glycine) followed by any sequence of four residues and by a glycine, a lysine, and a polar residue (serine or threonine). It is important to have a way to describe these patterns so that we can identify their presence in a protein of interest. For example, the ATP-binding pattern can be encoded as [AG]-X(4)-G-K-[ST].

Similarly, the pattern [LIVM]-[ST]-A-[STAG]-H-C defines an amino acid sequence starting with a hydrophobic residue, followed by a polar one and then an alanine, a small amino acid, a histidine, and a cysteine. This is characteristic of the region around the catalytic histidine of serine proteases.

Sequences and patterns are monodimensional data and hence easy to display. More complex is the task of visualizing biological structures. A three-dimensional effect in two dimensions can be achieved by a number of tricks, sometimes difficult to implement but easy to describe. For example, the depth of a three-dimensional structure can be rendered by diminishing the intensity of the parts farther from the observer (Figure 1.9) and by permitting the rotation of the molecule in real time.

Proteins are very complex objects formed by a large number of atoms and chemical bonds; the graphical representation of all of them can lead to incomprehensible pictures. We need to find ways to simplify the view—for example, by taking advantage of the fact that a protein is composed by repetitive elements (amino acids and secondary structure elements), as we can see in Figures 1.9 and 1.10, respectively.

1.4.1 PROTEIN ARCHITECTURE

A protein structure is the biological counterpart of a modern art masterpiece: Its beauty can make us breathless, even if we are not entirely sure about what the artist wanted to tell us. Over the years, hundreds of scientists have spent hours scratching their eyes and heads while looking at a computer screen trying to "understand." But what do they want to understand and what have we understood so far?

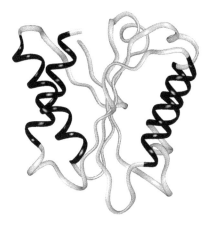

FIGURE 1.9 Visualization of a protein structure. The three-dimensional effect is obtained by hiding the lines on the back and by diminishing the intensity of the lines farther away from the observer.

The story goes that when Linus Pauling saw the three-dimensional structure of myoglobin, solved by Sir John Kendrew and coworkers in 1960, he was struck by its complexity, even though he had already predicted the existence and the geometry of α-helices. We can be sympathetic with his disconcertedness. After the beautiful regularity of the structure of a DNA molecule, a protein structure must have looked like a confused and disordered ensemble of atoms; certainly, for a protein, the mere structural information is not sufficient for understanding its function.

Nonetheless, proteins are reasonably regular objects that can be analyzed and classified in terms of their architecture and, sometimes, their details. It is impossible to summarize in a few paragraphs what we know about protein structures, and many books discuss this fascinating subject. Therefore, we will just mention a few facts that should already be known and that we will need in the next chapters.

Proteins are made by amino acids. Amino acids have different properties. Some are hydrophobic (Met, Phe, Ile, Leu, Val, Ala, Pro, Trp), some are polar (Gly, Tyr, Cys, Ser, Thr, Asn, Gln), and some are charged negatively (Asp, Glu) or positively (His, Arg, Lys). Some amino acids have peculiar behaviors: Cysteine can make disulfide bridges, glycine has no side chain, and proline is an immino acid.

With the exclusion of the latter amino acid, the main chain of an amino acid defines two solid angles, the angle ϕ (around the N–Cα bond) and the angle ψ (around the Cα–C bond) (Figure 1.11). When two amino acids condense to form a peptide bond, they give rise to another angle, ω, around the peptide bond (i.e., the bond between the C of one amino acid and the N of the subsequent one in the chain). This bond has a partial double-bond character; therefore, its rotational degree of freedom is limited and the bond is usually planar (i.e., the ω angle is usually 180°

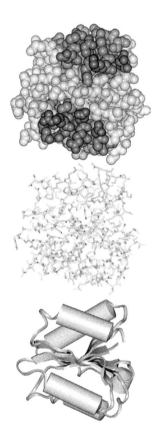

FIGURE 1.10 Different levels of complexity in displaying a protein structure. In the top picture, all nonhydrogen atoms are displayed as spheres; in the middle, all atoms are displayed as vertices of broken lines representing the bonds; the bottom picture shows a ribbon following the direction of the main chain. Compare these images with Figure 1.9, where the same protein is displayed using a ribbon connecting the α-carbons of its amino acids.

[*trans* conformation]). The other two angles, ϕ and ψ, can vary; however, for sterical reasons, some values are energetically unfavored and thereby rarely observed (Figure 1.12).

Figure 1.12 shows the so-called *Ramachandran plot*, a graph illustrating the energetically favorable combinations of the ϕ and ψ angles. The darker the region is the more favorable is the energy. The region indicated with α_L is generally disallowed because of the steric hindrance of the C_β atom with the oxygen of the main chain. Of course, this is not the case when the C_β is not there (i.e., for glycines). The regions indicated with β and α_R correspond to the angles formed by the polypeptide

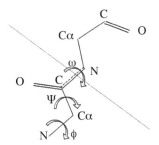

FIGURE 1.11 The polypeptide chain conformation is given by the three angles φ, ψ, and ω.

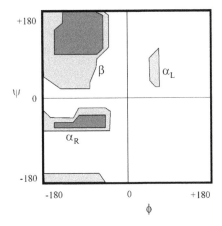

FIGURE 1.12 The Ramachandran plot. Phi-angle values are reported on the x-axis and ψ values on the y-axis. Shaded regions are energetically favored.

when it is folded as the most frequently observed secondary structures: the α-helices and β-strands (the elements forming β-sheets).

Linus Pauling predicted the existence of the α-helix before the determination of any protein structure. It is a repetitive structure, perfectly suited to saturate all hydrogen bonds of the main chain: The C=O group of one amino acid is hydrogen bonded to the N–H group of the amino acid four residues downstream in the chain. The geometry of α-helices is well known: Each amino acid translates 1.5 Å along the helix axis and rotates about 100° with respect to the previous residue; therefore, the helix pace is 3.6 residues.

Pauling also predicted the existence of β-sheets. In this case, hydrogen bonds are all saturated, but this time the hydrogen-bonded groups belong to different strands. Strands can run parallel (i.e., in the same direction) or antiparallel (in opposite directions). β-sheets can be made by only parallel strands (called parallel sheets), only antiparallel strands (antiparallel sheets), or both parallel and antiparallel

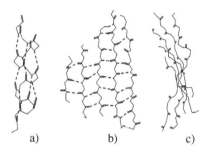

a)   b)   c)

FIGURE 1.13 The structures of (a) α-helix; (b–c) antiparallel β-sheet. The sheet in (c) is the same as in (b) rotated by about 90° to illustrate the fact that a sheet is generally not flat.

strands (mixed sheets). Note that β-sheets are not flat but usually are bent, as shown in Figure 1.13.

Both α-helices and β-strands are extended structures; the globular shape of most proteins is achieved by reversing the direction of the polypeptide chain, a role played by loops and turns.

Given the two secondary structure elements, we can assemble proteins made only of α-helices (α-class), only β-sheets (β-class), or both. In the latter case, α-helices and β-sheets can be separated in the structure (giving rise to the α+β-class) or packed together (α/β-class). Finally, proteins can have a low content of secondary structural elements. All these classes are observed in nature and the classification that we described forms the basis of almost all methods for the structural classification of proteins (e.g., the SCOP and CATH databases mentioned earlier).

Examples of proteins belonging to these different structural classes are shown in Figure 1.14. They include enzymes and structural and regulatory proteins. Some of them bind a cofactor and we know the structure of the complex; some bind metal ions.

REFERENCES

Historical Contributions

Three papers describe the discovery of the DNA double helix. Watson and Crick report the construction of the double helix model on the basis of diffraction experiments, which are described in Franklin and Gosling and Wilkins et al.:

Franklin, R., and R. G. Gosling. 1953. Molecular configuration in sodium thermonucleate. *Nature* 4356:740–741.

Watson, G. D., and F. H. Crick. 1953. Molecular structure of nucleic acids: A structure for deoxyribose nucleic acid. *Nature* 4356:737–738.

Wilkins, M. H. F., A. R. Stokes, and H. R. Wilson. 1953. Molecular structure of deoxypentose nucleic acids. *Nature* 4356:738–740.

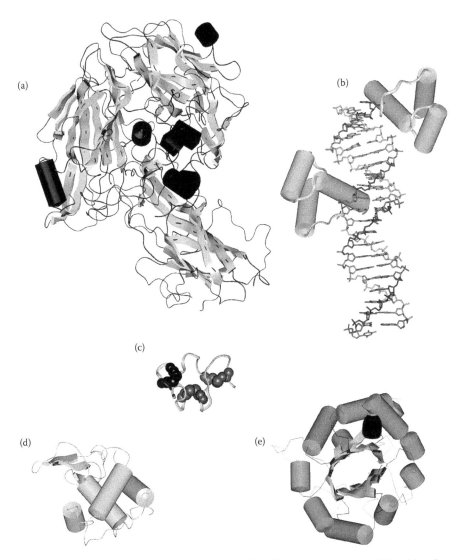

FIGURE 1.14 Examples of protein structures. (a) a β-protein, a domain of the rhinovirus capside protein (PDB ID: 4RHV); (b) An all α-protein, a DNA binding homeodomain (PDB ID: 3HDD); (c) a toxin, with low secondary structure content (PDB ID: 1AJJ); (d) an α+β-protein, human lysozyme (PDB ID: 1LZR) (e) an α/β-protein, triose phosphate isomerase (PDB ID: 1HTI).

In 1968 Jim Watson published a book called *The Double Helix* (New York: Atheneum). The subtitle read: *A Personal Account of the Discovery of the Structure of DNA*. Indeed, the point of view was really personal, and a few years later Anne Sayre, wife of a well-known crystallographer and a friend of Rosalind Franklin, published another book on the same subject called

Rosalind Franklin and DNA (New York: Norton and Co.). The book tries to tell the story from the point of view of Rosalind Franklin, who had passed away years before. Anne Sayre writes that she was initially puzzled by some apparently innocent mistakes in Watson's book, such as the use of Rosy—a nickname never used by the scientist or her friends—and, most of all, the chapter in which Watson wondered how could "Rosy" had looked without glasses—a very strange thought since Rosalind Franklin never needed to wear glasses.

Sayre was motivated by her view that we should beware of stories in which there are discrepancies with known facts. Therefore, she looked for (and found) other mistakes in Watson's book. Sayre's aim was to re-establish the truth on Rosalind Franklin and her role in the discovery of the structure of DNA. Nevertheless, reading her book and Watson's is very instructive and can contribute to understanding the historical period of a scientific discovery that represents a landmark in modern biology.

A prediction of the existence of α-helices and β-sheets was published in 1951:

Pauling, L., R. B. Corey, and H. R. Branson. 1951. The structure of proteins: Two hydrogen-bonded helical configurations of the polypeptide chain. *Proceedings of the National Academy of Sciences of the United States of America* 37:205–211.

Suggestions for Further Reading

At the beginning of each solar year, the scientific journal *Nucleic Acids Research* publishes a volume entirely dedicated to databases. In this volume, scientists in charge of the most important databases (GenBank, EMBL, SwissProt, PIR, PDB, Prosite, ProDom, SCOP, CATH, etc.) describe the state of the art and all new features of their systems.

Protein structures are beautifully described in many books:

Branden, C., and J. Tooze. 1993. *Introduction to protein structures.* Bologna: Zanichelli.

Lesk, A. 2001. *Introduction to protein architecture.* Oxford: Oxford University Press.

Perutz, M. F. 1992. *Protein structure: New approaches to disease and therapy.* New York: W. H. Freeman.

PROBLEMS

1. On which human chromosome is the β-globin gene located?
2. Find at least one disease associated with mutations on the human Y chromosome and report its OMIM id.
3. Find the chromosomal location of the gene homologous to human β-globin in rat, chimp, mouse, and chicken

4. The known nuclear targeting sequences are generally basic, but there seems to be no clear common denominator among all the known sequences, although some consensus sequence patterns have been proposed. One possible pattern is the following:

 two adjacent basic amino acids (Arg or Lys)

 a spacer region of any 10 residues

 at least three basic residues (Arg or Lys) in the five positions after the spacer region

 Write a Prosite-style pattern describing this pattern.

5. Find the PDB entry 1CKL and answer the following questions:
   a. Which protein is it?
   b. At which resolution has the structure been solved?
   c. What are the R and Rfree values?
   d. From which organism does the protein come?
   e. To which chromosome does the globin gene belong?

6. Search for the SRY gene. Where is it? How long is the corresponding protein? Are there known polymorphisms in its coding region? Can you find the pattern described in exercise 1.4 in this sequence? If so, at which position is it? What can be said about the protein?

7. Find the PDB entry 1NVL
   a. How many structures are in the file?
   b. Why?
   c. The protein binds a ligand: which one?
   d. At which pH was the structure solved?
   e. What is the secondary structure of this protein? Report the limits of β-strands and α-helices.
   f. Find the SCOP and CATH classification of the protein. Is there any difference? If so, can you discuss why a difference exists?

8. Which amino acids of the protein contact the ligand?

9. The human SNP rs10626313 is
   a. Located in a region coding for which gene?
   b. On which chromosome?
   c. Which type of SNP?

10. The most serious complication of cystic fibrosis is respiratory tract infection by the ubiquitous bacterium *Pseudomonas aeruginosa*. Has the genome of the bacterium been completely sequenced? Does it have a sensory transduction system for chemotaxis? Can you find which gene(s) is (are) involved in the chemotaxis process of this bacterium?

2 Genome Sequence Analysis

GLOSSARY

Algorithm: a set of operations necessary to carry out a task; the word derives from a Persian mathematician, Abu Jàfar Mohammed ibn Mûsâ al-Khowârizmî, who wrote a book on arithmetic around 800 A.D.

Artificial neural networks: algorithms capable of optimizing their parameters so as to reproduce an observed relationship between input and output data; after parameter optimization, network can be used for predicting the output

Eukaryote: organism with nucleated cells

Internet: large set of computers connected around the world

LINE: long interspersed nuclear sequences; long DNA regions (>1000 bp) repeated in the genome

Markov model: a sequence of states in which the probability of a state only depends on the previous one; named after Andrei Andreyevich Markov (1856–1922), a mathematician who studied poems and other texts as random sequences

ORF: open reading frame; a DNA sequence that does not contain a stop codon in at least one of the six possible reading frames

Primer walking: a DNA sequencing method in which already sequenced regions are used to design the primers for the next round of sequencing

Prokaryote: organism with non-nucleated cells

Sensitivity of a prediction method: the ability of a predictive method to detect a condition when, in fact, it is present

Shotgun: a DNA sequencing method in which the region to be sequenced is randomly fragmented, giving short redundant fragments that partially overlap

SINE: short interspersed nuclear sequences; short DNA regions (<1000 bp) repeated in the genome

Site-specific scoring matrices: matrices containing in each cell a value related to the probability that the base (or the amino acid) corresponding to the row occurs in the position corresponding to the column

Specificity of a prediction method: the ability of a predictive method to detect that a condition is not present when, in fact, it is not present

Splicing: processing of the newly transcribed RNA molecule, which excises the
introns, joins adjacent exons, and produces a molecule of messenger RNA
(mRNA)

Synonymous codons: codons encoding for the same amino acid

WWW: World Wide Web; network of documents connected by the Internet

2.1 BASIC CONCEPTS

When genome analysis is mentioned, we all tend to think about the human genome.
Even if *Homo sapiens* is only one of the many species whose genome has been
sequenced, there is no doubt that our expectations are all focused on the information
that we can gather about ourselves and our species. But, what is this information?

The precise knowledge of the sequence of the three billion nucleotides that make
up human DNA, besides being only theoretically possible because of intraspecie
variations, is not particularly relevant per se. Its importance is based on the fact that
we believe that we can use it to understand the properties of the various elements
inside a cell, opening the road to the targeted intervention, for example, in pathological processes.

The impact of genomic sequencing in biology is not only related to the amount
of available data, but also to its quality. A fundamental property of a genome is its
"completeness." In other words, once we know the sequence of the genome of an
organism, we also know that all that is needed by the organism must be encoded in
its genome—for example, all the enzymes needed for its metabolism, all the signaling proteins, all the proteins with a structural function, etc. The first consequence
is, for example, that the involvement of a given enzyme in a metabolic process can
not only be confirmed by its presence in the genome (the only possibility in the case
of a partially sequenced genome) but also excluded by its absence.

Therefore, given a complete genomic sequence, we need to localize the "useful"
regions, find the exact positions of the genes, derive the sequence of the encoded
molecules, understand the mechanism of their expression, discover their function,
and, finally, try to interfere specifically with the latter.

This is certainly not an easy task. More than 97% of the human genome does
not contain genes, but is made of pseudogenes, retropseudogenes, satellite regions,
minisatellites, microsatellites, transposons, retrotransposons, viral vestigials, etc.
Coding regions are interspersed in this large amount of so-called selfish DNA. Hence,
the first step in analyzing a genome requires "fishing" for the interesting regions.
This is worse than looking for a needle in a straw pile because at least the needle
is different from the straw, while, in the genome, we are looking for a thin straw in
a straw pile!

A more appropriate metaphor would be that we are attempting to identify the
paragraphs (the genes) of this book if they were randomly distributed in a 5,000-
page book (the genome), with the further complication that they would not be
contiguous, but interrupted by noninformative pieces of text often longer than the
paragraphs (the introns).

The aim of this chapter is briefly to describe the tools for extracting the sequence of a gene product from a genome sequence. In the following chapters, we shall discuss the bioinformatics tools that allow us to predict the function of these products.

2.2 GENOME SEQUENCING

Eukaryotes and *prokaryotes* genomes are different under many aspects critical for our analysis. The former are much bigger than the latter (the human genome is about 5,000 times longer than the largest bacterial genome), have a much lower gene density, contain long repetitive regions, and contain genes interrupted by introns. It follows that the techniques for sequencing and analyzing the genomes of eukaryotes and prokaryotes, and their respective difficulties, are very different, as we shall see. Nevertheless, all the genomes must first be reduced into fragments of "reasonable" dimensions and subcloned in order to be sequenced.

Biotechnology has been able to produce artificial vectors capable of stably maintaining DNA inserts up to 200,000 base pairs long that can be used to subclone regions of the genome. We need to know the correspondence between the cloned regions and their original position on a chromosome. This correspondence is called the physical map and can be obtained by different experimental methods. After this step, the sequence of the large fragments can be obtained by direct methods (*primer walking*) or by *shotgun*.

In direct methods, the sequence of a known fragment is used to design primers for sequencing the next contiguous piece of DNA. In this case, the reconstruction of the entire DNA sequence of the region is relatively straightforward.

In the shotgun method, all the clones are enzymatically or mechanically fragmented in random positions and subcloned. The process is such that the subcloned fragments are redundant and partially overlapping. The reconstruction of the "parent" clone is obtained by comparing all subclone sequences, finding the overlapping regions, and joining the sequences together in the correct order. At the end, direct sequencing strategies can be used, if needed, to finish the sequence.

Whichever the method, all sequences are checked for possible cross-contaminations coming from the reagents used in the experiments. This is obtained by comparing the obtained sequences with those of the cloning vectors, bacteriophage, and other micro-organisms. Subsequently, the sequences of eukaryotic genomes are screened for the presence of repetitive sequences that are usually of two types: satellite regions (generally found in specific zones of the chromosomes) or repetitive sequences randomly placed all over the DNA, called *SINE* (short interspersed nuclear sequences) or *LINE* (long interspersed nuclear sequences). A classic example of SINE is given by the Alu elements, which are present more than 300,000 times in the human genome (accounting for about 5% of the entire genome).

Although all these procedures are theoretically simple, none of them is easy in practice, if only because of the massive amount of data to be analyzed. However, especially for eukaryotic genomes, the real problems begin after these preliminary steps have been completed.

Where are the genes whose translation in amino acids (Table 2.1) will give us the sequence of the proteins the function of which is the aim of all our efforts?

TABLE 2.1
The Standard Genetic Code

UUU Phe	UCU Ser	UAU Tyr	UGU Cys
UUC Phe	UCC Ser	UAC Tyr	UGC Cys
UUA Leu	UCA Ser	UAA Stop	UGA Stop
UUG Leu	UCG Ser	UAG Stop	UGG Trp
CUU Leu	CCU Pro	CAU His	CGU Arg
CUC Leu	CCC Pro	CAC His	CGC Arg
CUA Leu	CCA Pro	CAA Gln	CGA Arg
CUG Leu	CCG Pro	CAG Gln	CGG Arg
AUU Ile	GCU Thr	GAU AsN	GGU Ser
AUC Ile	GCC Thr	GAC AsN	GGC Ser
AUA Ile	GCA Thr	GAA Lys	GGA Arg
AUG Met/Start	GCG Thr	GAG Lys	GGG Arg
GUU Val	ACU Ala	AAU Asp	AGU Gly
GUC Val	ACC Ala	AAC Asp	AGC Gly
GUA Val	ACA Ala	AAA Glu	AGA Gly
GUG Val	ACG Ala	AAG Glu	AGG Gly

2.3 FINDING THE GENES

A region of DNA not containing a termination codon is called open reading frame (*ORF*). In prokaryotic genomes a necessary (but not sufficient) condition for a DNA region to be a gene is the presence of a start codon (usually ATG) followed by an ORF and by one of the three termination codons (usually UAA, UAG, or UGA).

In eukaryotic genomes, almost all genes are formed by exons, each one of which is an ORF separated by noncoding regions, called introns, containing a variable number of nucleotides, from seven to many thousands. Usually, the junctions between exons and introns and introns and exons contain the base pairs AG and GT, respectively. Searching for genes in prokaryotes is clearly easier than in eukaryotes and, in first approximation, can rely upon the identification of ORFs.

The probability of randomly finding an ORF N-codon long, preceded by an ATG and followed by a termination codon, is the product of the probability of finding an ATG (1/64) by the probability of finding N coding codons ($(61/64)^N$) by the probability of finding one of the three stop codons (3/64):

$$P = (1/64) * (61/64)^N * (3/64)$$

N	Probability of randomly finding an N-codon ORF, starting with ATG and ending with a stop codon	Expected number of N-codon ORFs over 5 million base pairs (about the length of E. coli genome)
1	0.07%	3,500
10	0.045%	2,200
100	0.0006%	30
1000	10^{-22}%	Almost 0

We can assume that ORFs longer than a certain number of codons (e.g., 100) correspond to genes. Other parameters can be taken into account, such as the amino acid composition of the putative protein and the preferential usage of some codons among those codifying for the same amino acid.

Finding the genes in eukaryotic genomes is much more difficult. Let us review what we know about eukaryotic genes:

- Generally, there is a promoter upstream and a polyadenilation site downstream, although the specific sequences corresponding to them are not very well characterized.
- Introns are generally flanked by the AG and GT base pairs, but there is no constraint about their spacing
- An intron can contain any number of bases without recognizable features (for example, it can contain stop codons in any reading frame).
- We expect that the frequency of each codon in an exon is compatible with that expected for the specific organism.
- The analysis of gene sequences has revealed that exons and introns tend to have a periodicity in their sequence of nucleotides; the period is three for exons and two for introns.
- It has been noticed that context effects exist; in exons, some codons tend to be more frequently next to each other. Indeed, the frequency of hexa-nucleotides is one of the parameters that can be used to evaluate the probability that a given sequence codes for a protein.

All of these observations can be combined and used to develop methods for predicting the likelihood that a genome region is coding, as we will discuss in the next paragraphs. However, one property distinguishes coding regions from introns or intragenic regions: They are coding. This observation, far from being trivial, has some obvious and very important consequences.

First, the protein sequence coded by the exons could already be known. If one or more of the exons is reasonably long, a similarity search in a protein database might allow the identification of the final product. A comparison between the genomic sequence and its product could therefore lead to the identification of exons and introns.

Even if we do not know the sequence of the product of a gene, the sequence of a protein of the same family from a different source organism might be available. This can lead to the identification of the gene if the similarity between the products encoded by the genes of the two organisms is sufficiently high.

Fragments of the coding regions could already have been sequenced as EST (see the previous chapter) and therefore be present in the EST database. Although it might not allow the identification of the complete gene, the identification that an EST is present in the region is nonetheless a proof that a gene does exist in the region.

As we mentioned, derived databases contain sequence patterns related to specific functions. Once again, if a known functional pattern is found in an ORF, our confidence in its being a gene increases.

Finally, functionally important regions tend to be conserved during evolution. Therefore, comparing regions of the human genome with corresponding ones from another organism (such as mouse, for example) can lead to the identification of more conserved regions (most probably exons) from other, less well-conserved (all the rest) regions, even when we know nothing about either of the two genomic regions.

The features described here form the basis of available methods to identify eukaryotic genes that can be based on statistical or evolutionary considerations. Most of the time, they provide us with a table listing the putative coding regions and the associated probability that the identification is correct.

Let us briefly survey statistical methods for finding genes that take into account the properties of coding regions described here. Methods based on sequence similarity will be discussed in the next chapter.

2.4 STATISTICAL METHODS TO SEARCH FOR GENES

2.4.1 SITE-SPECIFIC SCORING MATRICES

Let us assume that we have a number of different sequences sharing a given property (e.g., the N bases upstream and downstream from a *splicing* site). We can describe the sequences using a matrix with $(2 * N)$ rows and four columns, where $2 * N$ is the number of positions and four is the number of possibilities (bases) in each position. The cells of the matrix contain a score reflecting the frequency with which the base corresponding to the row is observed in the position corresponding to the column. If the number of observations is sufficiently high, these frequency values approximate the probability of finding the base in the corresponding position.

Once this so-called site-specific matrix has been obtained, the search for a splicing site can be obtained by sliding the query sequence with respect to the matrix and multiplying the probabilities that each base would be observed in the corresponding position if it were part of the region around the splicing site. A high value of the product indicates a high probability that the sequence is part of the splicing region. The following table shows an artificial example of a *site-specific scoring matrix* with $N = 4$:

	1	2	3	4	5	6	7	8
A	0.30	0.50	0.10	0.45	0.65	0.32	0.12	0.20
T	0.20	0.15	0.05	0.20	0.05	0.25	0.30	0.44
C	0.45	0.20	0.80	0.30	0.25	0.40	0.20	0.30
G	0.05	0.15	0.05	0.05	0.05	0.03	0.58	0.26

The probability that the sequence GGTCACAACGTTAGG has properties similar to those of the sequences used to derive the matrix is the product:

$$0.05 * 0.15 * 0.05 * 0.30 * 0.65 * 0.40 * 0.12 * 0.20 = 7.02 * 10^{-7}$$

This assumes that the first G corresponds to the first position of the table (Example 1 in Figure 2.1). If instead we want to test the hypothesis that the second G of the sequence corresponds to the first position in the matrix, we obtain:

$$0.05 * 0.15 * 0.80 * 0.45 * 0.25 * 0.32 * 0.12 * 0.30 = 6.8 * 10^{-6}$$

(Example 2 in Figure 2.1), and so on.

The sequence AACCCACTA has a probability of:

$$0.30 * 0.50 * 0.80 * 0.30 * 0.25 * 0.32 * 0.20 * 0.44 = 2.53 * 10^{-4}$$

(Example 3), which is higher than the value computed for the previous two. But, is this probability sufficiently high to make us conclude that it represents the site described by the matrix? We can compare the value $1.65 * 10^{-4}$ with the value expected for a random sequence. If we assume that the nucleotide composition of a random sequence is 25% A, 25% T, 25% C, and 25%G, we should compare our computed values with:

$$0.25 * 0.25 * 0.25 * 0.25 * 0.25 * 0.25 * 0.25 * 0.25 = 1.53 * 10^{-5}$$

Under our random composition hypothesis, the sequence AACCCACTA has a significant probability of belonging to the class of sequences used to derive the matrix.

Probabilities are always lower than one; therefore, their product rapidly becomes very small as the number of positions increases. This can cause problems in our calculation. A practical way of facing the issue is to use the so-called log-odd ratio. We compute the logarithm (usually base 2) of the ratio between the observed frequencies of occurrence and the expected one. The logarithm of a product is the sum of the logarithms of the factors; hence, another advantage of the log-odd ratio is that we can sum the values instead of multiplying them. Going back to the previous example, by dividing each value of the matrix in Figure 2.1 by 0.25 and calculating its logarithm (base 2), we obtain the matrix in Figure 2.2.

If the observed frequency for a base in a position is higher (lower) than expected by chance, the ratio between the two is greater (smaller) than one, and the logarithm is greater (smaller) than zero. In the last table, in fact, a negative (positive) value in a cell indicates that the base of the row is present less (more) frequently in the position corresponding to the column than is expected for a random sequence.

Let us compute again the score of the two sequences GGTCACAAC and GTCA-CAACG using the log-odd matrix. For the first one we obtain:

$$-2.32 - 0.74 - 2.32 + 0.26 + 1.38 + 0.68 - 1.06 - 0.32 = -4.44$$

while, for the second:

$$-2.32 - 0.74 + 1.68 + 0.85 + 0.00 + 0.36 - 1.06 + 0.26 = -0.97$$

	1	2	3	4	5	6	7	8
A	0.30	0.50	0.10	0.45	0.65	0.32	0.12	0.20
T	0.20	0.15	0.05	0.20	0.05	0.25	0.30	0.44
C	0.45	0.20	0.80	0.30	0.25	0.40	0.20	0.30
G	0.05	0.15	0.05	0.05	0.05	0.03	0.58	0.26
Case 1 =	G	G	T	C	A	C	A	A
$7.02 * 10^{-7}$	0.05 *	0.15 *	0.05 *	0.30 *	0.65 *	0.40 *	0.12 *	0.20
Case 2 =	G	T	C	A	C	A	A	C
$6.8 * 10^{-6}$	0.05 *	0.15 *	0.80 *	0.45 *	0.25 *	0.32 *	0.12 *	0.30
Case 3 =	A	A	C	C	C	A	C	T
$2.53 * 10^{-4}$	0.30 *	0.50 *	0.80 *	0.30 *	0.25 *	0.32 *	0.20 *	0.44

FIGURE 2.1 An example of computing a score using a site-specific scoring matrix.

	1	2	3	4	5	6	7	8
A	0.26	1.00	-1.32	0.85	1.38	0.36	-1.06	-0.32
T	-0.32	-0.74	-2.32	-0.32	-2.32	0.00	0.26	0.82
C	0.85	-0.32	1.68	0.26	0.00	0.68	-0.32	0.26
G	-2.32	-0.74	-2.32	-2.32	-2.32	-3.06	1.21	0.06
Case 1	G	G	T	C	A	C	A	A
= -4.44	- 2.32	- 0.74	- 2.32	+ 0.26	+ 1.38	+ 0.68	- 1.06	- 0.32
Case 2	G	T	C	A	C	A	A	C
= -0.97	- 2.32	- 0.74	+ 1.68	+ 0.85	+ 0.00	+ 0.36	- 1.06	+ 0.26
Case 3	A	A	C	C	C	A	C	T
= 4.06	+ 0.26	+ 1.00	+ 1.68	+ 0.26	+ 0.00	+ 0.36	- 0.32	+ 0.82

FIGURE 2.2 Example of a log-odd site-specific scoring matrix. This matrix has been derived from the one in Figure 2.1 by assuming a random distribution where the four bases are uniformly represented in the sequence.

Both values are negative, since in most positions the sequences contain bases with a low probability of being in the region described by the matrix. On the other hand, the score of the sequence AACCCACTA is:

$$0.26 + 1.00 + 1.68 + 0.26 + 0.00 + 0.36 - 0.32 + 0.82 = 4.06$$

The frequency table that we used in our example does not have any cell containing zero; that is, we assumed that each base is present in each position at least once. We did so to avoid a problem: The logarithm of zero is $-\infty$ and this would make it impossible to compute the sum of the values. What can we do in real cases when a zero is present in the cells of the matrix? We simply add one to each of the cells of the matrix before taking its log-odd (method of the pseudocounts) as if at least one A, one T, one C, and one G were present in each position. The method of the pseudocounts is used very often, if not always, when we compute the logarithm of frequency values. It also has the added advantage that the probability of observing a base in a given position is never zero. This takes into account the fact that we might not have observed the base because of our limited sampling of the possible sequences and not because it is impossible to observe.

2.4.2 ARTIFICIAL NEURAL NETWORKS

Site-specific matrices evaluate the probability that the observed frequency of the bases in a given sequence is consistent with that expected for the specific region. This is a deterministic calculation and all the positions carry the same weight.

But, how can we be sure that all the positions are equally relevant for our final prediction? If this is not the case, how can we evaluate the relative weight of each position? We could see whether the values in each of the positions correlate with the property that we want to predict. For example, we can see how well they perform when used to predict the property in a set of known examples. This is an ideal application for artificial intelligence algorithms such as *artificial neural networks*.

Artificial neural networks are a class of algorithms modeled on biological neural networks. In biology, neurons receive exciting or inhibiting stimuli from other neurons and respond by passing the signal along whenever the stimulus exceeds a certain threshold. There are many different types of neural networks and this is not the right place to discuss the different underlying theories. We will use a simple example of one of the methods (called "supervised learning") to illustrate the basic concepts.

Let us take again the set containing the junctions between exons and introns and a different set containing approximately the same number of sequences of the same length that we know are not junctions. The first set is called "positive" the second "negative." Now, let us encode our sequences numerically.

There are different ways to do this. For example, we can number the DNA bases from one to four (say A = 1, T = 2, C = 3, G = 4), but this could create problems because it introduces a relationship between the bases (G is four times A, 2 is half of G, etc.) Usually, we use 4 bits to identify the bases (1,0,0,0), (0,1,0,0), etc., which

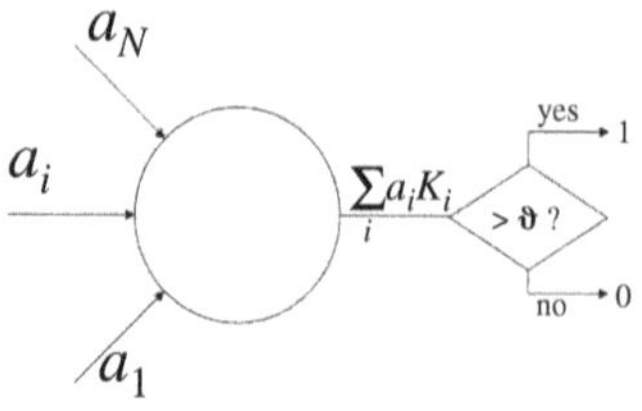

FIGURE 2.3 Schematic representation of an artificial neural network. The network will give the value 1 (0) if the result of the overall summation is greater (smaller) than the threshold θ.

is a waste of memory (2 bits are sufficient to identify four objects), but an advantage because each base is "equidistant" from any other.

Now we need an *algorithm* that will perform operations on our sequences. In what is called the training phase, the algorithm will be provided with the sequences as a series of binary numbers $a_1, a_2,..., a_i,..., a_N$ and with the expected result for each of the sequences (for example 1 for the "positive" set and 0 for the "negative" one). The neural network algorithm is able to modify its internal parameters and "learn" from the examples to give the desired answer.

In a very easy case, the algorithm could just calculate the following value (Figure 2.3):

$$X = \sum_i a_i K_i$$

and modify iteratively the K_i values (during the training phase) in order to obtain the expected result for as many cases as possible. We can select a threshold value θ and require that X is greater than θ when the positive cases are given as input and lower in all other cases.

Let us put sequences aside and illustrate the method with a simple numerical example:

	a_1	a_2	Expected
Example 1	1	0.3	Yes
Example 2	1	1	Yes
Example 3	0	0.8	No
Example 4	0.5	0.4	No

This problem has different solutions (for example, $K_1 = 1$, $K_2 = 0.5$, and a threshold θ = 1). Therefore:

example 1: $a_1 \times K_1 + a_2 \times K_2 = 1 \times 1 + 0.3 \times 0.5 = 1.15 \rightarrow$ yes (because greater than 1)

example 2: $a_1 \times K_1 + a_2 \times K_2 = 1 \times 1 + 1 \times 0.5 = 1.5 \rightarrow$ yes (because greater than 1)

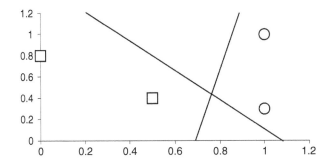

FIGURE 2.4 Geometric representation of an artificial neural network.

example 3: $a_1 \times K_1 + a_2 \times K_2 = 0 \times 1 + 0.8 \times 0.5 = 0.40 \rightarrow$ no (because less than 1)

example 4: $a_1 \times K_1 + a_2 \times K_2 = 0.5 \times + 0.4 \times 0.5 = 0.7 \rightarrow$ no (because less than 1)

This last example is also useful to introduce a different way of thinking about artificial neural networks (Figure 2.4). a_1 and a_2 can be seen as the coordinates of points in a plane. The task is to find a line (i.e., the two parameters that define a straight line) able to separate the points with coordinates, (1,0.3) and (1,1), whose expected result is "yes" from the points (0,0.8) and (0.5,0.4) whose expected result is "no."

Any of the straight lines shown in Figure 2.4 can discriminate between the two sets of points. If we train the network with more examples, the line can be moved (i.e., the network weights can be changed) so as to maintain the highest number of examples with expected value "no" separated from those with expected value "yes."

Let us now look at another example:

a_1	a_2	Expected
0	0	No
0	1	Yes
1	0	Yes
1	1	No

It can be easily verified that it is impossible to find a pair of K_1 and K_2 values able to give the expected output in all four cases.

The solution to this problem is to design a more complex neural network—that is, a network with more modifiable parameters—for example, using a hidden layer of neurons, as shown in Figure 2.5. In a graphical representation this means that more than one line or a higher order polynomial is needed to separate the two sets of points (Figure 2.6).

Other automatic learning methods can be used to address the mapping problem—for example, support vector machines. In this case, the details are different, but the underlying reasoning is not very different. We want to "teach" a computer program

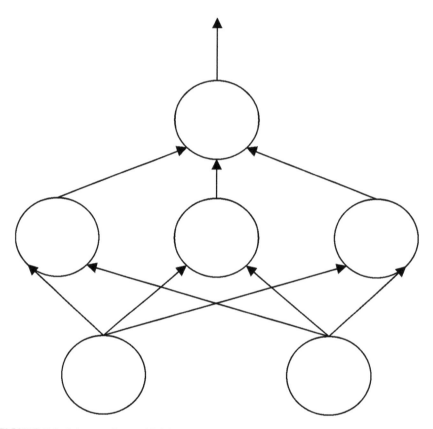

FIGURE 2.5 Scheme of an artificial neural network with a hidden layer.

to correctly map the input to the output by adjusting its parameters on the basis of a training set for which the mapping is known. The optimized parameters are subsequently used for predicting the output in cases where it is not known.

2.4.3 MARKOV MODELS AND HIDDEN MARKOV MODELS

When discussing site-specific matrices, we assumed that the probability of having a G in second position is, for example, 0.15 (because it had been observed 15% of the time in known cases), independently of the preceding base. In other words, we are assuming that the probability of having a G in second position is the same when preceded by an A, a T, a C, or a G.

However, for example, an acceptor splicing site is identified by the dinucleotide AG. Therefore, the probability of a base being at the beginning of an intron does depend upon the previous base, and this effect cannot be taken into account by a simple site-specific matrix. The statistical processes where the occurrence of an event depends upon the preceding ones are called Markov processes.

If we want to model our splicing site as a Markov process, we can calculate the frequencies of the doublets of bases in the region of interest (called region I). We

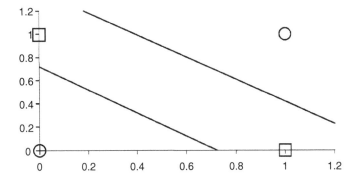

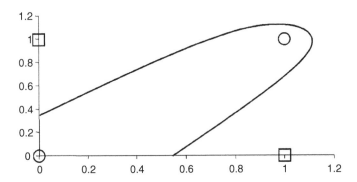

FIGURE 2.6 Geometrical representations of an artificial neural network with two input values and one hidden layer.

therefore use Table 2.2a, where the first row contains the probability that, in the region described by the matrix, an A is found after an A, a T, a C, or a G.

Let us now assume, for the sake of simplicity, that in the remainder of the sequence, which we will call region II, the probability of finding any nucleotide is always 0.25. The matrix describing region II is, therefore shown in Table 2.2b.

TABLE 2.2
Frequency of Doublets in Region I (a) and II (b)

	(a)					(b)			
	A	T	C	G		A	T	C	G
A	0.24	0.30	0.14	0.32	A	0.25	0.25	0.25	0.25
T	0.22	0.32	0.10	0.36	T	0.25	0.25	0.25	0.25
C	0.50	0.10	0.38	0.02	C	0.25	0.25	0.25	0.25
G	0.12	0.30	0.22	0.36	G	0.25	0.25	0.25	0.25

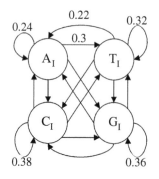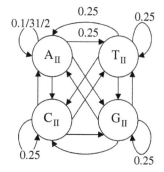

FIGURE 2.7 Markov models.

TABLE 2.3
Complexity of Some Genomes

	Average Number of Genes per 100,000 Bases	Average Number of Introns per Gene
Escherichia coli	87	0
Saccharomyces cerevisiae	52	0
Caenorhabditis elegans	22	3
Homo sapiens	5	6

We can depict the content of the two matrices in graphical form as shown in Figure 2.7, where each arrow corresponds to a probability value taken from the preceding matrices. For clarity, only some probability values are shown in the figure. As you can see, a CC- and CG-rich region is more likely to be found in region I than in region II.

How can we use these matrices to predict whether a sequence with unknown properties is more similar to region I than to region II and vice versa? We can compute the ratio between the probability of occurrence of a base (an A, for example) given the preceding base (for example, a T) in region II and compare with the probability of the same event in region I; that is, we can compute the ratio 0.22/0.25 = 0.88, take the logarithm of the result, and obtain a new scoring matrix:

	A	T	C	G
A	−0.058894	0.263034	−0.836501	0.356144
T	−0.184425	0.356144	−1.321928	0.526069
C	1	−1.321928	0.604071	−3.643856
G	−1.058894	0.263034	−0.184425	0.526069

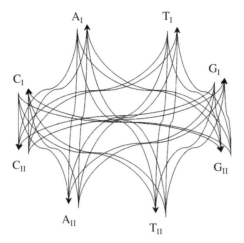

FIGURE 2.8 A hidden Markov model.

The matrix contains positive values for pairs observed more frequently in region II than in region I and it can be used to score a new sequence for its probability of belonging to one of the two regions.

If we are able also to compute the transition probability that a base in region I is followed by a base in region II and vice versa, we can use a *Markov model* that takes both regions into account (hidden Markov model, also known as HMM), as shown in the scheme of Figure 2.8, where the transitions within each of the regions (shown in the previous figure) are omitted. The transition probabilities are estimated heuristically by analyzing known sequences, similarly to what is done to derive site-specific matrices.

Markov models will be discussed more thoroughly in the chapter devoted to protein sequence alignments. Here we can say that a sequence of bases can be represented by several paths in the graph, according to whether we consider each base to belong to region I or region II. Each path is associated with a probability derived from the product of the transition probabilities of the steps in the path. The path with the highest probability will provide us with the most likely alignment of the new sequence with the sets of sequences used to derive the Markov model.

To give an example, in Figure 2.9 we show a simplified model with only two bases, each one with only two states. The model could have been generated by an alignment such as:

← I →	← II →
ATTA	TATA
TTAT	ATTA
AAAT	ATAT
....	....
TAAT	ATTT
TTAA	ATAT

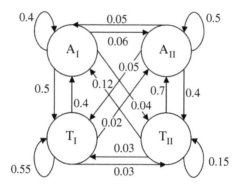

FIGURE 2.9 Example of a simple hidden Markov model.

Let us concentrate on the sequence TTAT. What is the probability that it belongs to zone I or zone II or that it is at the junction between the two regions?

TTAT could result from one of the following paths:

$T_IT_IA_IT_I$	$T_IT_{II}A_IT_I$	$T_{II}T_IA_IT_I$	$T_{II}T_{II}A_IT_I$
$T_IT_IA_IT_{II}$	$T_IT_{II}A_IT_{II}$	$T_{II}T_IA_IT_{II}$	$T_{II}T_{II}A_IT_{II}$
$T_IT_IA_{II}T_I$	$T_IT_{II}A_{II}T_I$	$T_{II}T_IA_{II}T_I$	$T_{II}T_{II}A_{II}T_I$
$T_IT_IA_{II}T_{II}$	$T_IT_{II}A_{II}T_{II}$	$T_{II}T_IA_{II}T_{II}$	$T_{II}T_{II}A_{II}T_{II}$

Each of them has a different probability of occurrence since the probability of each doublet (let us say AT) depends on the region to which the two components (A and T) belong. Therefore, we can calculate the probabilities associated with each path in our example:

$T_IT_IA_IT_I = 1.1 * 10^{-1}$	$T_IT_{II}A_IT_I = 1.8 * 10^{-3}$	$T_{II}T_IA_IT_I = 6.0 * 10^{-3}$	$T_{II}T_{II}A_IT_I = 9.0 * 10^{-3}$
$T_IT_IA_IT_{II} = 8.8 * 10^{-3}$	$T_IT_{II}A_IT_{II} = 1.4 * 10^{-4}$	$T_{II}T_IA_IT_{II} = 4.8 * 10^{-4}$	$T_{II}T_{II}A_IT_{II} = 7.2 * 10^{-4}$
$T_IT_IA_{II}T_I = 5.5 * 10^{-4}$	$T_IT_{II}A_{II}T_I = 1.0 * 10^{-3}$	$T_{II}T_IA_{II}T_I = 3.0 * 10^{-5}$	$T_{II}T_{II}A_{II}T_I = 5.2 * 10^{-3}$
$T_IT_IA_{II}T_{II} = 1.4 * 10^{-4}$	$T_IT_{II}A_{II}T_{II} = 8.4 * 10^{-3}$	$T_{II}T_IA_{II}T_{II} = 2.4 * 10^{-4}$	$T_{II}T_{II}A_{II}T_{II} = 4.2 * 10^{-2}$

From this, we can say that it is more likely that our sequence belongs to zone I since the path with the highest probability is $T_IT_IA_IT_I$ and can be aligned as shown in Figure 2.10.

The difference between a Markov model and a hidden Markov model should now be clear. In a Markov model, each base is unambiguously represented by only one state. In an HMM, each base is represented by more than one possible state (for example, A_I and A_{II}). Each sequence will be associated to the probability that it is generated by the model, but we cannot know in advance whether one of our bases (let us say A) will be considered as belonging to region I (A_I) or to region II (A_{II}) until we have computed the final result. This is why we call the model "hidden" and we need to decode it by calculating the probability assigned to each different path.

```
      I     II

    ATTA  TATA

    TTAT  ATTA

    AAAT  ATAT

    ..........  ..........

    TAAT  ATTT

    TTAA  ATAT

      TTAT
```

FIGURE 2.10 Possible alignment of the sequence TTAT to the multiple alignment used to generate the HMM shown in Figure 2.9.

Note that there are more efficient algorithms to decode the path than the exhaustive calculation that we used in our example.

An HMM can be as complex as we wish and therefore can be used to describe the structure of a gene, taking into account all the characteristics described at the beginning of this chapter. It can include probabilities derived by the frequencies of codons, dinucleotides, exanucleotides, etc., in exons and introns and it can include the frequency of the various bases upstream and downstream from a splicing site or in the polyadenilation site. Even the probability that an exon is evolutionarily related to a region of a known protein can be taken into account. Each base will have a different probability of being part of one of the gene elements and a different probability of being at the border between different gene elements. An HMM describing a gene can be represented by the scheme in Figure 2.11 (which is not very different from those used in real life).

2.4.4 Levels of Reliability

A prediction method—that is, a method aimed at predicting a property—can be very sensitive and can recognize a large fraction of the true cases (independently from how many times it incorrectly predicts the property in the negative cases) or very specific (i.e., it rarely assigns the property to elements that do not have it, but it does not necessarily recognize a large fraction of true cases). Let us assume that we have a method to predict whether a dinucleotide AG is a splicing signal and a set of cases for which the answer is known. Then we can compute (Table 2.4):

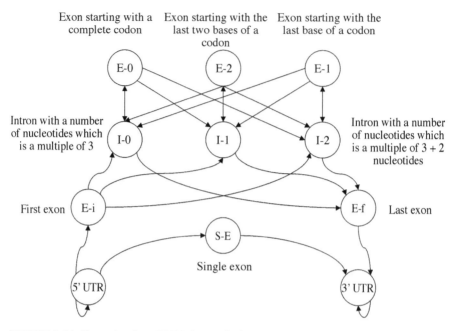

Exon starting with a   Exon starting with the   Exon starting with the
complete codon   last two bases of a   last base of a codon
codon

FIGURE 2.11 Example of an HMM for predicting the presence of a gene in a eukaryotic genome.

number of times the method predicts that an AG is a splicing site and is correct = TP (true positive)

number of times the method predicts that an AG is a splicing site and is incorrect = FP (false positive)

number of times the method predicts an AG is not a splicing site and is correct = TN (true negative)

number of times the method predicts an AG is not a splicing site and is incorrect = FN (false negative)

TABLE 2.4
Exemplifying the Parameters TP, FP, FN, and TN

		Has It Been Predicted as a Splicing Site?	
		Yes	No
Is it a splicing site?	Yes	TP	FN
	No	FP	TN

A method predicting that any AG is a splicing site, independently from anything else, will detect all the true splicing sites (no FN) and hence would be very sensible, but would also be quite useless since it would not be specific at all. On the other hand, a method predicting that none of the AG is a splicing site (no FP) would never mistakenly identify a nonsplicing site as one and therefore would be very specific but not sensitive at all.

The *sensitivity* = TP/(TP + FN) of a method represents the percentage of positive answers that the method correctly predicts. Its *specificity* = TP/(TP + FP) is the fraction of times in which the method correctly identifies positive cases. At present, the sensitivity and specificity of available methods for exon identification in eukaryotic genomes can reach almost 80%, but research in this field is quite active and these values tend to change on a daily basis.

For a better comparison among the different methods, it is more useful to have a single parameter depending on all the four values defined in Table 2.3 rather than taking into account sensitivity and specificity separately. One of the parameters used for this purpose is the correlation coefficient, defined as:

$$CC = \frac{TP \times TN - FP \times FN}{\sqrt{(TP + FP) \times (TP + FN) \times (TN + FP) \times TN + FN)}}$$

It is easy to see that the correlation coefficient assumes the value 1 only when all cases (positive and negative) are correctly predicted.

Sensitivity, specificity, and correlation coefficient are computed using a set of known test cases, as we mentioned, and it is important to stress that their values are reliable only when the set of test cases is correctly balanced, containing a similar fraction of positive and negative examples. Another way to analyze the predictive power of a method is the ROC (receiver operator curve), discussed in problem 9.

2.5 COMPARATIVE GENOMICS

Genome analysis is extremely useful in our efforts to understand the biology of an organism. Comparative genomics is even more essential. As we mentioned, it can help in identifying coding regions and in assigning them a function and it is also instrumental for the identification of untranslated regions involved in gene regulation. Furthermore, it can help address issues related to the mechanism of evolution of different species and of their physiology.

The availability of more than one genome, for example, enables us to discover how many and which proteins are necessary and sufficient for providing the basic functions of one organism. How much this set differs in size and type in different organisms also has important practical outcomes. For example, comparative genomics enabled us to find out that the number of genes involved in functions such as transcription and translation is essentially constant throughout evolution, while some specific functions, such amino acids biosynthesis, metabolism, transport, and regulation, are carried out by a different number of genes in different organisms (the number seems to be approximately proportional to the genome size).

An interesting case is that of the transporter proteins in prokaryotes. Compound transport is crucial to bacteria, since their cytoplasmic membrane is the only barrier between the cytosol (where all the physiological processes take place) and the outside world, where conditions can be very harsh. For example the pathogen *Helicobacter pylori* survives in the very acidic medium of the gastrointestinal tract.

Complete genomic sequencing can be used to identify proteins involved in membrane trafficking so that their behavior can be analyzed and used for practical applications. One example for all occurs in the genome of the bacterium responsible for tuberculosis. There are nine types of the enzyme ATPase, each specific for different divalent cations: why so much redundancy? Could it be linked to the need of the pathogen for a specific resistance? Could it be exploited to cure the disease?

Several organisms can use different strategies and all are very interesting to study and understand. The bacterium responsible for influenza (*Haemophilus influenzae*) and that responsible for pneumonia (*Mycoplasma pneumoniae*) affect the respiratory apparatus, but the first has at least 13 proteins for amino acid import, while the latter has only 3 with a broader specificity. These different strategies are probably due to evolution (i.e., to their different needs to adapt to their ecological niche). On the other hand, the knowledge of these "idiosyncrasies" of the pathogens could help in treating these two diseases in a focused way. As a general rule, transport strategies in an organism can influence the survival rate in different environments; therefore, an analysis of the genes involved in the process can lead to a targeted search for antimicrobial drugs and also to the design of particular growing media in which the micro-organisms can be propagated and studied.

Comparative genomics is also revealing an unexpected high frequency of horizontal gene transfer—that is, transfer of genetic material between organisms. Besides the theoretical interest of how and why such a transfer happens, there is a practical interest in studying this phenomenon since it seems to be correlated with pathogenicity. Sometimes gene loss is also linked to pathogenicity. For example, *H. pylori*, *Mycobacterium tuberculosis*, *Mycoplasma genitalium*, and *M. pneumoniae* have all lost the genes for proofreading nucleic acids during replication. Hence, they are more variable and able to escape the host immune system more efficiently.

2.6 A VIRTUAL WINDOW ON GENOMES: THE WORLD WIDE WEB

Genomes made their first appearance in databases at the beginning of the 1980s. First came the genomes of viruses, phage, and organelles. More or less at the same time, molecular biology developed efficient and fast sequencing techniques. At the moment of writing this book more than 2,000 genomes have been completely sequenced. Most of them (about 1,000) are viral genomes, about 800 are organelles (mostly mitochondria), about 300 are bacterial, a couple of dozen are archeobacteria, and the others are eukaryotic genomes. These figures are growing exponentially and the numbers given here could be archived as outdated as soon as they are written. How can this be handled?

This question brings forward another general subject, although we will discuss it very superficially: What is the impact of the World Wide Web on bioinformatics?

At the beginning of the 1990s, a bioinformatics group subscribed to databases and the magnetic tapes (does anyone remember them?) containing the known sequences were delivered on a monthly or trimonthly basis. Those were read and memorized on local computers and used for database searches, statistical analyses, etc. The size of the data was really limited. At the beginning of the 1980s, I published an analysis of the sequences of proteins of known structure only armed with a pocket calculator, a (nonelectronic) notebook, and a pencil! A few years later CD-ROM appeared and at least the delivery of the data was made easier and lighter. Soon after, it became possible to update the local copy of the databases between one release and another by connecting to one of the database centers and downloading the new sequences via the *Internet*.

It was not going to last for long. In the meantime, the World Wide Web (*WWW*) exploded and, thanks to this, it is now possible to use the databases directly at their location, without the need of storing a local copy. The main advantage does not only lie in this latter aspect, but rather in the fact that up-to-date data can be accessed in real time.

The flexibility and dynamics of the World Wide Web make it an indispensable tool in every research laboratory, but it also has problems. Sites tend to move, disappear, and be born again somewhere else with different interfaces. Any list of bioinformatics Web sites becomes obsolete even more quickly than the statistics of genomes. That is the reason why there are no lists of sites in this book (except for a few stable and reliable ones mentioned in the last chapter) and no instructions for using available Web tools. The reader needs to consult the related Web pages and read the instructions there since the only way to manage the WWW is through the WWW itself.

REFERENCES

Historical Contributions

The human genome sequence has been published at the same time by a consortium of public research institutes in the scientific journal *Nature* and by a consortium of private research institutes in the scientific journal *Science*:

Lander, E. S., L. M. Linton, et al. 2001. Initial sequencing and analysis of the human genome. *Nature* 409:860–921.

Venter, J. C., M. D. Adams, et al. 2001. The sequence of the human genome. *Science* 5507:1304–1351.

Suggestions for Further Reading

The methods for gene hunting are many and continuously growing. A description of the underlying general principles can be found in Baxevanis, A. D. 2001. Predictive methods using DNA sequences. In *Bioinformatics: A*

Practical Guide to the Analysis of Genes and Proteins, Baxevanis, A. D. and Oullette, B. F., eds., pp. 233–252. New York: Wiley–Liss, Inc.
The following book contains a detailed mathematical description of the probabilistic methods applied to biological problems: Durbin, R., S. R. Eddy, A. Krogh, and G. Mitchison. 1998. *Biological Sequence Analysis*. Cambridge: Cambridge University Press.

PROBLEMS

1. How many codons should an ORF contain to have a probability lower than 0.01% of being observed by chance? How many such random ORFs do we expect in the *H. pylori* genome (assume a random distribution of codons in the genome)?
2. Derive a log-odd matrix from the following alignment:

Alignment

A	T	T	T	T	G	C	G	T	A	C
T	A	A	T	A	T	A	G	T	A	G
A	T	T	T	G	A	C	G	T	C	A
A	T	T	T	T	G	C	G	T	A	C
T	A	T	T	A	A	A	G	T	A	G
A	T	T	C	G	A	A	G	T	C	G
A	T	A	T	T	G	C	G	T	G	C
T	A	A	T	A	T	A	G	T	A	G
A	T	A	C	G	A	A	G	T	C	G

3. Compute the sensitivity, specificity, and correlation coefficient of a hypothetical method that is expected to give a positive score when the property is present, and negative otherwise.

Score	Present	Absent
Positive	18	1
Negative	14	92
Totals	32	93

4. Find the present sensitivity and specificity of gene-finding methods.
5. Analyze the region of the *P. aeruginosa* genome containing the putative PA0017 gene. Is its sequence similar to other known or putative genes in other organisms? Does this make it more or less likely to be a real gene?
6. Look at the region containing the gene for carnitine O-acetyltransferase of the human genome and answer the following questions:
   How many putative exons are in the gene?
   How many alternative forms of the gene are transcribed?

How many have been experimentally observed?

7. Design (schematically) a neural network for discriminating between coding and noncoding ORFs on the basis of their sequence composition and length.

8. Describe a strategy for training your neural network. Which genes would you select for training and testing? How would you collect the information?

9. A prediction method can be used with different parameters selected by the user. The results are:

Threshold	Sensitivity	Specificity
≤1.2	0.939	0.123
1.4	0.909	0.281
1.6	0.800	0.400
1.8	0.700	0.550
2.0	0.500	0.650
2.2	0.485	0.773
2.4	0.394	0.843
2.6	0.333	0.891
2.8	0.333	0.896
≥3.0	0.250	0.920

Find the definition of the ROC, draw it for this method, and comment on the results.

10. The area under the ROC (AUC) is commonly used for evaluating the quality of a method. Compute the AUC for the ROC curve of problem 9 and say which threshold value you would suggest to use.

3 Protein Evolution

GLOSSARY

BLOSUM: a family of similarity matrices derived from well-conserved aligned regions in protein families (based on the BLOCKS database)

Bootstrap: in evolutionary tree construction, a technique to evaluate the reliability of the tree

PAM: point accepted mutation; measure of the evolutionary distance between sequences. Two sequences are at 1 PAM distance if there has been one accepted mutation for every 100 amino acids. It is also used to indicate a series of similarity matrices.

Phylogenetic tree: a method to depict and visualize evolutionary relationships between molecules or organisms; similar to a family genealogic tree

Protein family: group of evolutionarily related proteins sharing a common ancestor

Similarity matrices: scoring matrices containing a value related to the probability that each amino acid is replaced by each other during evolution

UPGMA: unweighted pair group method using arithmetic averages; an algorithm to build phylogenetic trees

3.1 BASIC CONCEPTS

Galileo wrote in his *Dialogue of the Maximum Systems*:

> But surpassing all stupendous inventions, what sublimity of mind was his who dreamed of finding means to communicate his deepest thoughts to any other person, though distant by mighty intervals of place and time! Of talking with those who are in India; of speaking to those who are not yet born and will not he born for a thousand or ten thousand years; and with what facility, by the different arrangements of 20 characters upon a page!

This citation could easily be applied to evolution and what can be achieved by the different arrangements of 20 compounds in a protein.

Evolution-selected molecules have many different roles in the life of an organism, just by arranging the same 20 amino acids in different order. Proteins catalyze chemical reactions, build up a cell and its substructures, react to external stimuli, transmit signals, bind nucleic acids, and, in doing so, regulate transcription and translation, etc. Our task is to decode the role of a protein starting from its particular combination of amino acids and hence *to associate one or more specific functions to proteins of known sequence*.

We need to be careful when using the word "function" because it has a quite broad meaning. It is relatively easy to define the function of an enzyme, but we are not made up only of enzymes. Enzymes are classified by a system of four digits: N_1, N_2, N_3, and N_4. The first digit (N_1) indicates the reaction catalyzed by the enzyme: 1 = oxidoreductase, 2 = transferase, 3 = hydrolase, 4= lyase, 5= isomerase, and 6 = ligase. The second (N_2) identifies the class of chemical group on which the enzyme acts. For example, hydrolases can break: 3.1 an ester, 3.2 a glycosidic, 3.3. an ether, 3.4 a peptide bond, etc. N_3 describes more specifically the enzymatic function; for example, 3.4.11 is an aminopeptidase, 3.4.13 is a dipeptidase, etc. The last number identifies the specific enzyme.

This classification is necessary but not sufficient to characterize an enzyme's function. The mechanism that a protein uses to perform its task and the nature and position of the groups involved are crucial. We need to understand the molecular mechanism of the function of a protein and hence its detailed three-dimensional "shape." Every enzyme catalyzes a chemical reaction by placing functional groups in precise relative positions so that they are able to facilitate the reaction. These are the groups that we would like to interfere with in order to modulate the reaction or to block the activity of the enzyme.

The structure assumed by a protein is also essential to elucidate the function of proteins other than enzymes. Two proteins or a protein and a nucleic acids molecule are able to interact specifically because their shapes are complementary and because the groups that come in contact upon complex formation can form energetically favorable interactions with each other. Therefore, the phrase written in italics earlier must be written in a more specific way: *to associate one or more specific functions at a molecular level to each protein of known sequence.*

Thus, the task of understanding the function of a protein is related to the knowledge of the structure that it assumes in the appropriate physiological environment.

By and large, the three-dimensional structure of a protein depends exclusively on its amino acid sequence. In an historical experiment, Anfinsen proved that a protein that has been chemically denatured (i.e., unfolded) goes back to its native structure if the denaturing agents are removed and the physiological conditions are restored. This implies that the amino acid sequence univocally determines the structure of a protein. As always in biology, there are some exceptions that we will not discuss.

In general, we can state that a "structural code" links the amino acid sequence of a protein to its three-dimensional structure. Cracking the code that relates sequence to structure is one of the most challenging (and fascinating) tasks of modern biology, but, after decades of efforts, we are still far from a general solution of the problem.

3.2 MOLECULAR EVOLUTION

If we are not able to associate a three-dimensional structure directly to an amino acid sequence in order to understand the molecular details of its function, why should we dedicate so much time and effort to determining and analyzing protein sequences?

Of course the knowledge of a protein sequence is a crucial step to determine its structure, but this fact alone does not justify the enormous investment of the scientific community in sequencing projects. A protein sequence carries great value because it can often be used to infer the protein's function, sometimes even at a molecular level. The underlying reason for this is that all natural proteins are a product of evolution.

Nucleic acid sequences undergo mutations, deletions, insertions, crossing-over, etc. All these variations have a direct effect on the coded proteins. If a protein sequence is conserved along evolution (i.e., is present in many different organisms, thus giving rise to a so-called *protein family*), it is reasonable to assume that it might have a similar function in all the organisms and, if the evolutionary distance is not very large, that it might have preserved the amino acids crucial for its function.

We will discuss to which extent this hypothesis is correct. However, within its limits, if we know the function and/or structure of a member of an evolutionary family, we can often predict the function of all the other members and even identify the important functional groups (for example, the catalytic site).

What are the necessary steps to achieve this?

First, we need to identify which proteins belong to the same family. These proteins evolved from the same ancestor by a set of accepted mutation events; hence, their amino acid sequences are likely to be more similar than expected for unrelated sequences. Next, we need to identify the matching residues within the family—the ones sharing the same functional and/or structural role in the different members of the family. For didactical reasons, we will invert the two steps and first discuss the methods to align two sequences belonging to the same family (i.e., to obtain the correspondence between amino acids sharing equivalent positions in the two proteins) and then those aimed at retrieving protein sequences belonging to the same family from protein sequence databases.

3.3 HOW TO ALIGN TWO SIMILAR SEQUENCES

The problem of aligning two protein sequences in the context of function assignment can be formulated as: Which of the possible correspondence (alignment) between the amino acids of the two proteins is more likely to reflect their evolutionary relationship?

It seems reasonable to assume that it is the alignment involving the lowest number of mutations. In first approximation, this implies finding the alignment that minimizes the difference (or maximizes the similarity) between the amino acid sequences of the two proteins. We are looking for the alignment that places the largest number of identical (or similar) amino acids of the two proteins in corresponding positions. Let us assume that we have two protein sequences and that we write their sequences in the first row and the first column of a matrix (as shown in Figure 3.1).

If the amino acid in position i of the horizontal sequence is equal to the one in position j of the vertical sequence, we fill the element i,j of the matrix. If the two sequences were identical, all the cells of the main diagonal (joining the upper left-hand corner with the lower right-hand corner) would be filled. This representation

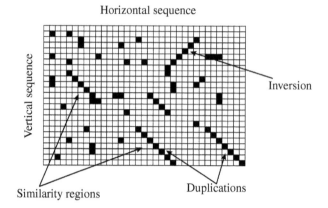

FIGURE 3.1 A dot plot.

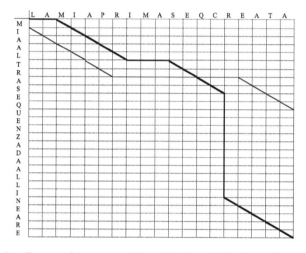

FIGURE 3.2 An alignment is represented by a line in a similarity matrix.

is called dot plot and makes it easy to identify regions of similarity (indicating identical subsequences), duplications (two or more parallel filled diagonal stretches indicate that one region of one protein is identical to more than one region of the second protein), and inversions (diagonals orthogonal to the main one). Any alignment of the two sequences is represented by a segmented line starting from the upper left-hand corner and ending in the lower right-hand corner.

The thin line of Figure 3.2 corresponds to the following alignment:

```
-LAMIAPRIMASEQCREATA----------------

MIAALTR---------ASEQUENZADAALLINEARE
```

The M of the vertical sequence does not correspond to any letter of the horizontal sequence, I corresponds to L, A to A, etc. Horizontal and vertical lines indicate insertion in the horizontal and vertical sequence, respectively.

Analogously, the thicker line corresponds to the alignment:

```
LAMIAPRIMASEQC------------REATA

--MIAAL---TRASEQUENZADAALLINEARE
```

If we assigned the value "1" to cells corresponding to identical amino acids and "0" to the others, the optimal alignment corresponds to the line that passes through more "1" or (which is equivalent) to the line for which the sum of the traversed cells (called the total score) is the highest, as shown in Figure 3.3.

Therefore, our first task is to devise an algorithm that, given a matrix, finds the path with the highest total score. Clearly, the elements of the matrix do not need to be just 0 and 1, but can contain any number. For example, as we shall see, they can be filled with a value related to the similarity between the amino acid of the row and that of the column, or with the probability that the two amino acids replaced each other during evolution.

In the alignment matrix, vertical or horizontal lines correspond to insertions or deletions. These are rarer events than substitutions. Therefore, it is intuitive, but will also be more evident later, that we should somehow penalize their appearance in

	L	A	M	I	A	P	R	I	M	A	S	E	Q	C	R	E	A	T	A
M	0	0	1	0	0	0	0	0	1	0	0	0	0	0	0	0	0	0	0
I	0	0	0	1	0	0	0	1	0	0	0	0	0	0	0	0	0	0	0
A	0	1	0	0	1	0	0	0	0	1	0	0	0	0	0	0	1	0	1
A	0	1	0	0	1	0	0	0	0	1	0	0	0	0	0	0	1	0	1
L	1	0	0	0	0	0	0	0	0	0	0	0	0	0	0	0	0	0	0
T	0	0	0	0	0	0	0	0	0	0	0	0	0	0	0	0	0	1	0
R	0	0	0	0	0	0	1	0	0	0	0	0	0	0	1	0	0	0	0
A	0	1	0	0	1	0	0	0	0	1	0	0	0	0	0	0	1	0	1
S	0	0	0	0	0	0	0	0	0	0	1	0	0	0	0	0	0	0	0
E	0	0	0	0	0	0	0	0	0	0	0	1	0	0	0	1	0	0	0
Q	0	0	0	0	0	0	0	0	0	0	0	0	1	0	0	0	0	0	0
U	0	0	0	0	0	0	0	0	0	0	0	0	0	0	0	0	0	0	0
E	0	0	0	0	0	0	0	0	0	0	0	1	0	0	0	1	0	0	0
N	0	0	0	0	0	0	0	0	0	0	0	0	0	0	0	0	0	0	0
Z	0	0	0	0	0	0	0	0	0	0	0	0	0	0	0	0	0	0	0
A	0	1	0	0	1	0	0	0	0	1	0	0	0	0	0	0	1	0	1
D	0	0	0	0	0	0	0	0	0	0	0	0	0	0	0	0	0	0	0
A	0	1	0	0	1	0	0	0	0	1	0	0	0	0	0	0	1	0	1
A	0	1	0	0	1	0	0	0	0	1	0	0	0	0	0	0	1	0	1
L	1	0	0	0	0	0	0	0	0	0	0	0	0	0	0	0	0	0	0
L	1	0	0	0	0	0	0	0	0	0	0	0	0	0	0	0	0	0	0
I	0	0	0	1	0	0	0	1	0	0	0	0	0	0	0	0	0	0	0
N	0	0	0	0	0	0	0	0	0	0	0	0	0	0	0	0	0	0	0
E	0	0	0	0	0	0	0	0	0	0	0	1	0	0	0	1	0	0	0
A	0	1	0	0	1	0	0	0	0	1	0	0	0	0	0	0	1	0	1
R	0	0	0	0	0	0	1	0	0	0	0	0	0	0	1	0	0	0	0
E	0	0	0	0	0	0	0	0	0	0	0	1	0	0	0	1	0	0	0

FIGURE 3.3 Example of optimal alignment between two sequences. Insertions and deletions are not included in the calculation of the path.

the alignment; otherwise, the algorithm might insert too many of them in order to maximize the total score. For example, the alignment corresponding to the line in Figure 3.3 is:

```
LAMIAP---RIMASEQ-----------CREATA

--MIA-ALTR--ASEQUENZADAALLIN--EARE
```

where there are five insertions/deletions.

To summarize, in order to obtain an alignment between two proteins, we need:

- a method to assign a score (a measure of the similarity between amino acids, to be used in place of the 0 and 1 values in the matrices)
- a strategy to penalize gaps (insertions/deletions)
- an alignment algorithm

3.4 SIMILARITY MATRICES

We have mentioned that we can fill the alignment matrix with values indicating identity or nonidentity of amino acids (1 or 0), as well as with values reflecting the similarity between two amino acids. These latter values are reported in *similarity matrices*: 20×20 matrices where each cell contains a value related to the similarity between the amino acid corresponding to the row and that corresponding to the column.

In theory, we could analyze the shape, type, and volume of each amino acid, deduce how similar it is to each of the other nineteen amino acids, and finally quantify these observations to derive the probability that a given amino acid can substitute another during evolution. In practice, it is more convenient and probably also more correct to use a statistical approach and assign values that reflect how often each amino acid is replaced by another in families of evolutionarily related (homologous) proteins.

For example, if we have many pair-wise alignments between similar proteins known to be homologous, we can calculate the frequency f_{ij} of occurrence of the amino acids i and j in corresponding positions in the alignments (i.e., the number of i and j pairs divided by the total number of aligned pairs). Next, we calculate the frequencies f_i and f_j with which the amino acids i and j are found in the sequences.

The ratio $f_{ij}/(f_i \bullet f_j)$ indicates how frequently the i and j residues are in corresponding position in the alignments with respect to what is expected by chance alone. The similarity matrices contain the $\log_2$ of this ratio for reasons similar to those discussed in Chapter 2 for site-specific matrices.

It should not have escaped the attention of the reader that we are using alignments to calculate the substitution matrix and that we need the matrix to generate an alignment. We must solve the problem of obtaining reliable manual alignments in order to deduce the matrices that we will use to align other sequences in more difficult cases. The main difference between the most commonly used matrices is indeed related to which alignments are used to generate them.

3.4.1 PAM MATRICES

The main idea behind *PAM* matrices is to use pairs of very similar sequences to derive the substitution frequencies and to extrapolate the values of the frequencies from them to what is expected for more divergent sequences.

PAM means "point accepted mutation." Two sequences are at 1 PAM distance if we can convert one into the other with an average of one accepted mutation every 100 amino acids. "Accepted" means that the mutation is not lethal for the organism and is accepted in the population. If we use many pairs of sequences at 1 PAM distance, we expect about 1% differences between each pair and we can derive the frequencies of substitution for each of the amino acid pairs. The probability of two independent events is the product of the two individual probabilities. Therefore, the substitution frequencies expected for sequences at a distance of 2 PAM are given by the product of the 1 PAM distance matrix by itself, a 3 PAM distance matrix is the product of the 1 PAM distance matrix by a 2 PAM matrix, and so on.

As sequence divergence increases, the probability that a single position undergoes more than one mutation also increases. Two sequences at 100 PAM distance do not have 100% of mutated amino acids, but rather a lower number because the amino acids can have been mutated more than once. The difference between the PAM number and the percentage of different amino acids increases with the distance. A PAM 80 matrix represents sequences with an average of 50% identical amino acids, sequences distant 250 PAM have about 20% identical amino acids, and so on. The PAM 250 matrix is shown in Figure 3.4.

	A	C	D	E	F	G	H	I	K	L	M	N	P	Q	R	S	T	V	W	Y
A	2																			
C	-2	12																		
D	0	-5	4																	
E	0	-5	3	4																
F	-4	-4	-6	-5	9															
G	1	-3	1	0	-5	5														
H	-1	-3	1	1	-2	-2	6													
I	-1	-2	-2	-2	1	-3	-2	5												
K	-1	-5	0	0	-5	-2	0	-2	5											
L	-2	-6	-4	-3	2	-4	-2	2	-3	6										
M	-1	-5	-3	-2	0	-3	-2	2	0	4	6									
N	0	-4	2	1	-4	0	2	-2	1	-3	-2	2								
P	1	-3	-1	-1	-5	-1	0	-2	-1	-3	-2	-1	6							
Q	0	-5	2	2	-5	-1	3	-2	1	-2	-1	1	0	4						
R	-2	-4	-1	-1	-4	-3	2	-2	3	-3	0	0	0	1	6					
S	1	0	0	0	-3	1	-1	-1	0	-3	-2	1	1	-1	0	2				
T	1	-2	0	0	-3	0	-1	0	0	-2	-1	0	0	-1	-1	1	3			
V	0	-2	-2	-2	-1	-1	-2	4	-2	2	-2	-2	-1	-2	-2	-1	0	4		
W	-6	-8	-7	-7	0	-7	-3	-5	-3	-2	-4	-4	-6	-5	2	-2	-5	-6	17	
Y	-3	0	-4	-4	7	-5	0	-1	-4	-1	-2	-2	-5	-4	-4	-3	-3	-2	0	10

FIGURE 3.4 PAM250 matrix.

Some interesting properties of amino acids can be derived by looking at the matrix. For example, the substitution arginine–triptophan has almost the same probability of the arginine–lysine substitution. Can you explain why?

3.4.2 BLOSUM Matrices

We briefly described derived databases in Chapter 1. Among them we mentioned BLOCKS, a database storing the sequence alignments of the most conserved regions of protein families. These alignments are used to derive the *BLOSUM* matrices, with the same method used for the PAM matrices (i.e., by normalizing the calculated frequencies of amino acid substitutions by the frequencies of the amino acids in the alignment). Not all the sequences of the alignments are used to derive the BLOSUM matrix; sequences with a percentage of identity higher than a certain threshold are averaged and considered as just one. We will have a different matrix for each threshold. For example, BLOSUM62 matrix (Figure 3.5) is derived from an alignment after all sequences with more than 62% identities with any other sequence of the alignment have been averaged.

Therefore, while in the PAM series, the lower the value of the matrix is the higher is the sequence similarity of the alignments used to derive it, in the BLOSUM series a lower value indicates that more divergent sequences have been used. A PAM250 roughly corresponds to a BLOSUM20, a PAM80 to a BLOSUM50, a PAM110 to a BLOSUM60, and so on.

	A	C	D	E	F	G	H	I	K	L	M	N	P	Q	R	S	T	V	W	Y
A	4																			
C	0	9	-3																	
D	-2	-3	6																	
E	-1	-4	2	5																
F	-2	-2	-3	-3	6															
G	0	-3	-1	-2	-3	6														
H	-2	-3	-1	0	-1	-2	8													
I	-1	-1	-3	-3	0	-4	-3	4												
K	-1	-3	-1	1	-3	-2	-1	-3	5											
L	-1	-1	-4	-3	0	-4	-3	2	-2	4										
M	-1	-1	-3	-2	0	-3	-2	1	-1	2	5									
N	-2	-3	-1	0	-3	0	1	-3	0	-3	-2	6								
P	-1	-3	-1	-1	-4	-2	-2	-3	-1	-3	-2	-2	7							
Q	-1	-3	0	2	-3	-2	0	-3	1	-2	0	0	-1	5						
R	-1	-3	-2	0	-3	-2	0	-3	2	-2	-1	0	-2	1	5					
S	1	-1	0	0	-2	0	-1	-2	0	-2	-1	1	-1	0	-1	4				
T	0	-1	-1	-1	-2	-2	-2	-1	-1	-1	-1	0	-1	-1	-1	1	5			
V	0	-1	-3	-2	-1	-3	-3	3	-2	1	1	-3	-2	-2	-3	-2	0	4		
W	-3	-2	-4	-3	1	-2	-2	-3	-3	-2	-1	-4	-4	-2	-3	-3	-2	-3	11	
Y	-2	-2	-3	-2	3	-3	2	-1	-2	-1	-1	-2	-3	-1	-2	-2	-2	-1	2	7

FIGURE 3.5 BLOSUM62 matrix.

3.5 PENALTIES FOR INSERTIONS AND DELETIONS

The detection of similar regions in sequence alignments is important because they highlight amino acids important for the function and/or the structure of the protein family, but the information contained in divergent regions is also important. For example, regions exposed to solvent are more tolerant to insertions and deletions, so it is crucial to find them in an alignment.

All the matrices that we use are derived from alignments of very similar sequences, without insertions or deletions. Therefore, we need to treat them separately. These are less frequent events and therefore their occurrence in an alignment needs to be penalized.

Unfortunately, no theoretical treatment can help us to assign the penalty for insertions and deletions; their position and length are dependent upon the specific structural and functional properties of each protein family. We need to derive the penalty values empirically, by optimizing them on the basis of tests performed using pairs of proteins whose structure—and therefore amino acid correspondence—is known. Consequently, they depend upon the matrix and the algorithm used and each alignment program provides the users with suggested optimal penalty values.

Usually, we use two gap penalties. One is called the "opening" penalty for insertions and the other (usually lower than the former) is the "continuation" of an insertion or a deletion. Intuitively, this derives from the observation that a protein can tolerate insertions in only a limited set of positions (for example, on the surface and outside of secondary structure elements). Therefore, starting a new insertion or deletion is more risky and thus "costly" than elongating an existing insertion or deletion.

A useful practical suggestion to verify whether the used gap penalty values are not interfering with the quality of our alignment is to use values different from those suggested by the programs' authors, although not very distant from them (as we said, gap penalties are related to the matrix used to derive the score of the alignment). Regions of the alignment that are stable with respect to limited variations in gap penalties are more reliable than those whose alignment substantially changes as soon as penalty values are slightly modified.

3.6 THE ALIGNMENT ALGORITHM

In this section we will describe one alignment algorithm. More are available, but their basic principles are similar.

First, let us recall that we are looking for the path that maximizes the score "collected" in the alignment matrix. Let us assume that we are playing a game that consists in going from the START to the END positions in the playfield shown in Figure 3.6 through the path that allows us to collect the highest number of golden pots. We cannot go backward; that is, we are only allowed to proceed from left to right.

Let us consider the position marked by "A." There are only two ways to go from START to "A": using the path marked by the arrows 1 and 2 or that marked by the arrows 3 and 4. In the former case, we would collect three golden pots, in the latter, five. This implies that the maximum number of pots that we can collect going from

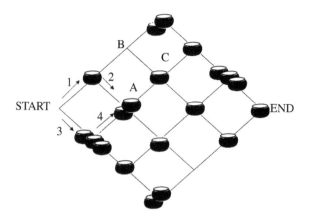

FIGURE 3.6 The golden pot game. The aim of the game is to collect the highest number of pots without going backward or passing twice through the same point.

START to END passing through point A is five plus the maximum number of pots that we can collect going from A to END. In other words, the path that we will choose to go from A to END does not depend on how we reached A. From the figure, it is also clear that the maximum number of pots that we can collect when going from START to B is one.

Let us now focus our attention on point C. We can get there by passing through A or B. What is the maximum number of pots that we can collect when going through C? At most, it is the maximum number of pots collected in A (five pots) plus the one in C (a maximum of six) or the maximum pots collected in B (one pot) plus the one in C (a maximum of two pots). It derives that the maximum number of pots that we can collect by going from START to C is six.

We can repeat this reasoning for each of the nodes and record, for each node, the maximum number of pots that we can collect from START to that node and, also, a pointer to the node we came from. This procedure is shown in Figure 3.7. It is clear that the maximum possible number of pots that we can collect going from START to END is 11 and the best path is the one that would lead us from END to START following our pointers backwards.

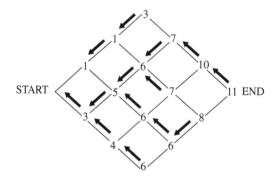

FIGURE 3.7 Solution to the game in Figure 3.6.

	H	E	A	G	A	W	G	H	E	E
P	-2	-1	-1	-2	-1	-4	-2	-2	-1	-1
A	-2	-1	-5	0	5	-3	0	-2	-1	-1
W	-3	-3	-3	-3	-3	15	-3	-3	-3	-3
H	10	0	-2	-2	2	-3	2	10	0	0
E	0	6	-1	-1	-3	-3	-3	0	6	6
A	-1	-1	5	5	0	-3	0	-1	-1	-1
E	0	6	-1	-1	-3	-3	-3	0	6	6

FIGURE 3.8 Scoring matrix for the alignment of the sequences HEAGAWGHEE and PAWHEAE.

Let us apply the same idea to protein sequence alignments. The only differences are that the values of the similarity matrices (corresponding to our golden pots) can also be negative (as if in the golden pot game we could lose pots in some nodes) and that we can insert or delete amino acids (i.e., "jump" over some nodes). In order to achieve a high score, an algorithm would jump positions with negative scores (i.e., use many insertions and deletions); this behavior should be penalized since insertions and deletions are not very frequent in homologous proteins and because they are difficult to place in a structural framework.

The first step of this procedure consists in building a score matrix where the cell (i,j) contains the similarity value between the amino acid in position i of the first sequence and the amino acid j of the second sequence taken from the chosen similarity matrix. Just as in the game before, the second step is to build a second matrix, called cumulative matrix, in which each element contains the maximum score that can be obtained when aligning the two sequences up to that point. A scoring matrix (that we will call A) is shown in Figure 3.8. It has been built using the similarity scores from a BLOSUM62 matrix.

Let us now compute the maximum score that we can achieve in a generic position (i,j) of the cumulative matrix that we call B. There are three ways to get to $B(i,j)$:

from $B(i-1,j-1)$ by summing the value of the cell $A(i,j)$ to the score obtained up to $A(i-1,j-1)$—that is, $B(i-1,j-1)$
from $B(i-1,j)$ by adding to the score obtained up to $A(i-1,j)$—that is, $B(i-1,j-1)$, the penalty for an insertion (which is a negative value)
from $B(i,j-1)$ by adding to the score obtained up to $A(i,j-1)$—that is, $B(i-1,j-1)$, the penalty for an insertion

The element $B(i,j)$ of the cumulative matrix should contain the maximum among these three values (Figure 3.9).

The algorithm fills a cell (i,j) using the content of the cells preceding it $(i-1$ and $j-1$ for each i and j value, including 1). Therefore, we need to build a matrix with one extra row and one extra column: row 0 and column 0. All the positions in each of these two represent insertions or deletions (since we do not align them with any amino acid). Hence, we will fill row 0 and column 0 starting from 0 and adding

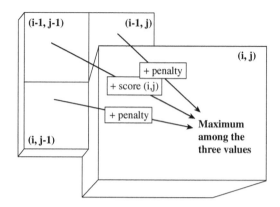

FIGURE 3.9 Calculating the score of a generic cell (*i,j*) in a cumulative matrix.

		H	E	A	G	A	W	G	H	E	E
	0	-8	-16	-24	-32	-40	-48	-56	-64	-72	-80
P	-8										
A	-16										
W	-24										
H	-32										
E	-40										
A	-48										
E	-56										

FIGURE 3.10 Filling row 0 and column 0 in a cumulative matrix.

the penalty value for insertion in each cell. For example, if the penalty is –8, row 0 and column 0 will be filled as shown in Figure 3.10. We can also choose not to penalize insertions and deletions at the beginning of the alignment and fill row 0 and column 0 with a value of 0.

The cell $B(1,1)$ will contain a value that is the maximum among the following:

−2 = the value of the cell $B(0,0)$, which is 0, plus the value of $A(1,1)$, which is −2

−16 = the value of $B(0,1)$ cell, which is −8, plus the penalty −8

−16 = the value of $B(1,0)$, which is −8, plus the penalty −8

Therefore, the value assigned to $B(1,1)$ will be −2. We will also record that we obtained this value coming from the $B(0,0)$ cell.

Let us fill another cell, $B(1,2)$, with the highest score among the following:

$B(0,1) + A(1,2) = -8 - 1 = -9$

$B(0,2) + penalty = -16 - 8 = -24$

$B(1,1) + penalty = -2 - 8 = -10$

$B(1,2)$ will contain −9 and a pointer to the $B(0,1)$ cell

The final cumulative matrix for this example is shown in Figure 3.11.

	H	E	A	G	A	W	G	H	E	E
	0	**-8**	**-16**	-24	-32	-40	-48	-56	-64	-72	-80
P	-8	-2	**-9**	**-17**	**-25**	**-33**	-41	-49	-57	-65	-73
A	-16	-10	-3	**-4**	**-12**	**-20**	-28	-36	-44	-52	-60
W	-24	-18	-11	-6	-7	-15	**-5**	**-13**	-21	-29	-37
H	-32	-14	-18	-13	-8	-9	-13	-7	**-3**	-11	-19
E	-40	-22	-8	-15	-15	-9	-12	-15	-7	**3**	-5
A	-48	-30	-15	-3	-11	-10	-12	-12	-15	**-5**	2
E	-56	-38	-23	-11	-6	-12	-13	-15	-12	-9	**1**

FIGURE 3.11 Cumulative matrix corresponding to the scoring matrix of Figure 3.8.

The value of the bottom rightmost cell is the highest score that can be achieved by any global alignment, given our similarity matrix and gap penalties. This score can be obtained if we include the $B(7,10)$ cell. In turn, the score of this latter cell is obtained if the previous one is $B(6,9)$ and so on. The cells in bold in Figure 3.11 indicate the path contributing to the highest score; thus, they are part of the optimal alignment. We can reconstruct the alignment starting from the bottom rightmost cell in bold and proceeding backward. Multiple cells in bold in one row or one column correspond to insertions in the first or second sequence, respectively. This simple example admits more than one solution; that is, more than one alignment has an optimal score, given our parameters (Figure 3.12).

H	E	A	G	A	W	G	H	E	-	E
-	-	P	-	A	W	-	H	E	A	E

H	E	A	G	A	W	G	H	E	-	E
-	P	-	-	A	W	-	H	E	A	E

H	E	A	G	A	W	G	H	E	-	E
-	P	A	-	-	W	-	H	E	A	E

FIGURE 3.12 Alignments obtained from the cumulative matrix of Figure 3.10.

3.7 MULTIPLE ALIGNMENTS

Amino acids that are essential for the function and/or the structure of a protein are under evolutionary pressure; therefore, they are conserved (or substituted by very similar residues). They can be identified in the alignment of sequences of homologous proteins. However, if two sequences are very close in evolution, most of their residues will have remained unchanged and it will be difficult to detect important residues among them. On the other hand, if two proteins are very distant in evolution a reliable alignment of their sequences will be much more difficult to obtain.

One way to solve this problem is to align the highest number of sequences of homologous proteins, whenever possible. Each pair of sequences will contain residues that are conserved because of evolutionary pressure and others conserved by chance, but the latter will be different for each pair. Consequently, when many sequences are aligned, the columns where all sequences contain similar or identical amino acids will stand out against the background and this will be instrumental in detecting the important regions of the proteins, as well as in identifying the exposed parts, where insertions and deletions tend to accumulate.

Professor Arthur Lesk says, "Two homologous sequences whisper, a multiple sequence alignment shouts." A "shouting" multiple alignment of a protein family is shown in Figure 3.13. Can you determine the amino acids crucial for the function and structure of the family? Can you identify the exposed regions of the proteins?

It is very likely that residues in positions 1, 8, 9, 10, 12, 13, 15, 16, and 19 have an important role, given their conservation within the family, and that the region 41–54 corresponds to an exposed region, probably a loop, since it includes insertions and deletions.

The conclusions that we derived from the alignment are correct, as can be appreciated by looking at Figure 3.14, where the structure of a member of the family is shown. The conserved residues are in the region that binds the flavin cofactor and the region 41–54 is indeed an exposed loop.

The preceding discussion should have convinced the reader of the importance of aligning families of proteins rather than sequence pairs. Unfortunately, the algorithms used for aligning pairs of sequences are not easily extensible to more than a few sequences. Present methods for multiple sequence alignments work progressively, aligning first a pair of sequences, then a third sequence to the alignment of the first two, and so on.

It is wiser to align first the two most similar sequences, to obtain a more reliable initial alignment. Most available multiple-sequence alignment methods align every possible sequence pair and calculate the respective score in order to decide in which order the sequences should be aligned. Note that, in this step, we only need to compute the optimal score and not the alignment, so it will be sufficient to record the value in the bottom rightmost cell of the alignment matrix and there will be no need to reconstruct the path that gave rise to it.

Some methods use sequence similarity to build a *phylogenetic tree*, which is a way to visualize the sequence similarity relationships between the protein sequences. A phylogenetic tree is similar to a genealogic tree (Figure 3.15), but it deals with species rather than with individuals (Figure 3.16). In a phylogenetic tree each species

Positions 1–20:

	1	2	3	4	5	6	7	8	9	10	11	12	13	14	15	16	17	18	19	20
FLAV_CLOBE	A	.	.	.	.	I	V	Y	W	S	G	T	G	N	T	E	K	M	A	E
CYSJ_THIRO	A	.	.	I	T	I	L	F	G	S	Q	T	G	N	A	K	A	V	A	E
CYSJ_ECOLI	A	.	.	I	T	I	I	S	A	S	Q	T	G	N	A	R	R	V	A	Q
NOS2_CHICK	A	.	K	V	T	V	I	Y	A	T	E	T	G	K	S	E	T	L	A	N
NOS2_ONCMY	A	.	.	T	V	L	Y	A	T	E	T	G	K	S	Q	Q	T	L	A	Q
NOS1_RABIT	A	.	K	A	T	I	L	Y	A	T	E	T	G	K	S	Q	A	Y	A	K
NOS3_HUMAN	A	.	K	A	T	I	L	Y	G	S	E	T	G	R	A	Q	S	Y	A	Q
NOS_RHOPR	A	.	K	A	T	I	L	F	A	T	E	T	G	K	S	E	M	Y	A	R
NOS_ANOST	A	.	K	A	T	V	L	Y	A	T	E	T	G	R	S	E	Q	Y	A	R
NOS_LYMST	A	.	K	C	S	I	F	Y	A	T	E	T	G	R	S	E	R	F	A	R
NCPR_HUMAN	A	.	N	I	I	V	F	Y	G	S	Q	T	G	T	A	E	E	F	A	N
NCPR_CANTR	A	.	N	T	L	L	L	F	G	S	Q	T	G	T	A	E	D	Y	A	N
NCPR_SCHPO	A	.	.	A	A	V	F	F	G	S	Q	T	G	T	A	E	D	F	A	Y
NCPR_YEAST	A	.	N	Y	L	V	L	Y	A	S	Q	T	G	T	A	E	D	Y	A	E
FLAV_DESSA	A	.	K	S	L	I	V	Y	G	S	T	T	G	N	T	E	T	A	A	E
FLAV_DESGI	A	.	K	A	L	I	V	Y	G	S	T	T	G	N	T	E	G	V	A	E
FLAW_DESGI	A	.	.	.	L	V	F	G	S	T	T	G	N	T		E	T	V	A	E
FLAV_DESDE	A	Q	.	K	V	L	I	V	F	G	S	T	T	G	N	T	E	S	I	A	A
FLAW_BACSU	A	.	K	I	L	L	V	Y	A	T	M	S	G	N	T	E	A	M	A	D
FLAV_BACSU	A	.	K	A	L	I	T	Y	A	S	M	S	G	N	T	E	D	I	A	F
FLAV_MEGEL	A	.	.	.	.	I	V	Y	W	S	G	T	G	N	T	E	A	M	A	N
FLAV_TREPA	A	.	K	V	A	V	I	F	W	S	G	T	G	H	T	E	T	M	A	R
FLAV_SYNY3	A	.	K	I	G	L	F	Y	G	T	Q	T	G	V	T	E	T	I	A	E
FLAV_SYNP7	A	.	K	I	G	L	F	Y	G	T	Q	T	G	V	T	Q	T	I	A	E
FLAV_ANASP	A	.	K	I	G	L	F	Y	G	T	Q	T	G	K	T	E	S	V	A	E
FLAV_SYNP2	A	.	K	I	G	L	F	F	G	T	Q	T	G	V	T	E	E	L	A	Q
FLAV_KLEPN	A	.	I	I	G	I	F	F	G	S	D	T	G	N	T	E	N	I	A	K
FLAV_HAEIN	A	.	I	V	G	L	F	Y	G	S	D	T	G	N	T	E	N	I	A	K
FLDA_HELPJ	A	.	K	I	G	I	F	F	G	T	D	S	G	N	A	E	A	I	A	E
FLAW_ECOLI	A	.	N	M	G	L	F	Y	G	S	S	T	C	Y	T	E	M	A	A	E
Consensus	A	.	.	.	.	.	.	Y	G	S	.	T	G	.	T	E	.	.	A	

Positions 21–40:

	21	22	23	24	25	26	27	28	29	30	31	32	33	34	35	36	37	38	39	40
FLAV_CLOBE	L	I	A	K	G	I	I	E	G	K	D	V	N	T	I	N	V	S	D	V
CYSJ_THIRO	Q	L	G	A	R	A	S	E	G	M	D	A	R	V	I	S	M	G	D	F
CYSJ_ECOLI	A	L	R	D	D	L	L	A	K	L	N	V	K	L	V	N	A	G	D	Y
NOS2_CHICK	S	L	C	S	L	F	S	.	A	F	N	T	K	I	L	C	M	D	E	Y
NOS2_ONCMY	R	L	N	S	M	L	N	.	A	F	N	S	R	L	L	C	M	E	E	Y
NOS1_RABIT	T	L	C	E	I	F	K	.	A	F	D	A	K	V	M	S	M	E	E	Y
NOS3_HUMAN	Q	L	G	R	L	F	R	.	A	F	D	P	R	V	L	C	M	D	E	Y
NOS_RHOPR	K	L	G	D	I	F	S	.	A	F	H	S	Q	V	L	S	M	E	D	Y
NOS_ANOST	Q	L	V	E	L	L	G	.	A	F	N	A	Q	I	Y	C	M	S	D	Y
NOS_LYMST	R	L	S	E	I	F	K	.	V	F	H	S	R	V	V	C	M	D	D	Y
NCPR_HUMAN	R	L	S	K	D	A	.	H	Y	G	.	M	R	G	M	S	A	D	P	E
NCPR_CANTR	K	L	S	R	E	L	H	S	F	G	.	L	K	T	M	V	A	D	F	A
NCPR_SCHPO	R	F	S	T	E	A	K	A	F	N	.	L	T	N	M	V	F	D	L	E
NCPR_YEAST	K	F	S	K	E	L	V	A	F	N	.	L	N	V	M	C	A	D	V	E
FLAV_DESSA	Y	V	A	E	A	F	E	N	E	I	D	V	E	L	L	K	N	V	T	D	V
FLAV_DESGI	A	I	A	K	T	L	N	S	G	M	E	T	T	V	V	N	A	A	D	V
FLAW_DESGI	V	V	A	K	V	L	E	E	G	M	A	V	D	L	K	N	A	T	K	V
FLAV_DESDE	K	L	E	E	L	I	A	A	G	H	E	V	T	L	L	N	A	A	D	I
FLAW_BACSU	L	I	E	K	G	L	Q	E	E	L	A	E	V	D	R	F	E	A	N	D	I
FLAV_BACSU	I	I	K	D	T	L	Q	E	E	L	D	I	D	C	V	E	I	N	D	M
FLAV_MEGEL	E	I	E	A	A	V	K	A	G	A	D	V	E	S	V	F	R	E	D	T
FLAV_TREPA	C	I	V	E	G	L	N	V	G	A	K	A	D	L	F	S	V	M	D	F
FLAV_SYNY3	L	I	Q	K	E	M	G	G	S	.	V	V	D	M	M	D	I	S	Q	A
FLAV_SYNP7	S	I	Q	Q	E	F	G	G	S	.	I	V	D	L	N	D	I	A	N	A
FLAV_ANASP	I	I	R	D	E	F	G	N	.	V	V	T	L	H	D	V	S	Q	A	
FLAV_SYNP2	A	I	Q	A	A	F	G	G	D	.	I	V	E	L	F	D	V	A	E	V
FLAV_KLEPN	M	I	Q	K	Q	L	G	K	.	V	A	D	V	H	D	I	A	K	S	
FLAV_HAEIN	Q	I	Q	K	Q	L	G	S	.	L	I	D	I	R	D	I	A	K	S	
FLDA_HELPJ	K	I	S	K	A	I	G	N	.	A	E	V	I	D	V	A	K	A		
FLAW_ECOLI	K	I	R	D	I	I	G	P	.	L	V	T	L	H	N	L	K	D	D	
Consensus	.	.	.	.	.	.	.	.	.	.	.	.	.	L	.	.	.	.	.	.

Positions 41–55:

	41	42	43	44	45	46	47	48	49	50	51	52	53	54	55
FLAV_CLOBE	N	.	.	.	.	.	.	.	.	.	I	D	E	L	L
CYSJ_THIRO	N	P	R	.	.	.	.	.	.	.	K	L	S		
CYSJ_ECOLI	K	F	K	.	.	.	.	.	.	.	Q	I	A		
NOS2_CHICK	N	I	S	.	.	.	.	.	.	.	D	L	E		
NOS2_ONCMY	N	F	S	.	.	.	.	.	.	.	D	M	E		
NOS1_RABIT	D	I	V	.	.	.	.	.	.	.	H	L	E		
NOS3_HUMAN	D	V	V	.	.	.	.	.	.	.	S	L	E		
NOS_RHOPR	D	M	S	.	.	.	.	.	.	.	K	I	E		
NOS_ANOST	D	I	S	.	.	.	.	.	.	.	S	I	E		
NOS_LYMST	A	V	E	.	.	.	.	.	.	.	T	L	E		
NCPR_HUMAN	E	Y	D	L	A	D	L	S	.	S	L	P	E	I	D
NCPR_CANTR	D	Y	.	.	.	D	F	E	.	N	F	G	D	I	T
NCPR_SCHPO	N	Y	.	.	.	D	L	T	.	D	L	D	N	F	D
NCPR_YEAST	N	Y	.	.	.	D	F	E	.	S	L	N	D	V	P
FLAV_DESSA	S	V	.	.	.	.	.	.	.	.	A	D	L	G	N
FLAV_DESGI	T	A	.	.	.	.	.	.	.	.	P	G	L	A	E
FLAW_DESGI	K	A	.	.	.	.	.	.	.	.	A	G	L	A	E
FLAV_DESDE	S	A	.	.	.	.	.	.	.	.	E	N	L	A	D
FLAW_BACSU	D	D	.	.	.	.	.	.	.	.	A	Q	L	F	T
FLAV_BACSU	D	.	.	.	.	.	.	.	.	.	A	S	C	L	T
FLAV_MEGEL	N	.	.	.	.	.	.	.	.	.	V	D	D	V	A
FLAV_TREPA	D	.	.	.	.	.	.	.	.	.	V	G	T	F	D
FLAV_SYNY3	D	.	.	.	.	.	.	.	.	.	V	D	D	F	R
FLAV_SYNP7	D	.	.	.	.	.	.	.	.	.	A	S	D	L	N
FLAV_ANASP	E	.	.	.	.	.	.	.	.	.	V	T	D	L	N
FLAV_SYNP2	D	.	.	.	.	.	.	.	.	.	I	E	A	L	R
FLAV_KLEPN	S	.	.	.	.	.	.	.	.	.	K	E	D	L	E
FLAV_HAEIN	S	.	.	.	.	.	.	.	.	.	K	E	D	I	E
FLDA_HELPJ	S	.	.	.	.	.	.	.	.	.	K	E	Q	F	N
FLAW_ECOLI	S	.	.	.	.	.	.	.	.	.	P	K	L	M	E
Consensus															

FIGURE 3.13 Multiple sequence alignment of a region of the protein family of flavodoxin from *Clostridium beijerinckii*. The row indicated with "consensus" contains the amino acids conserved in more than 66% of the sequences.

FIGURE 3.14 The structure of flavodoxin (PDB code: 3nll). The residues conserved in the alignment of Figure 3.13 are shown in black. The thick ribbon indicates the 41–54 region. The molecule shown as a set of spheres is the flavin.

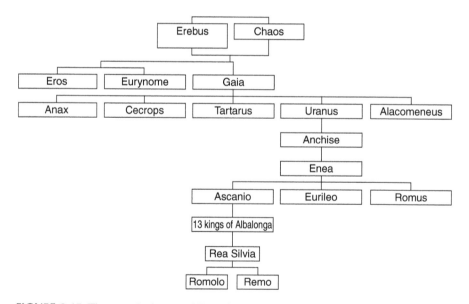

FIGURE 3.15 The genealogic tree of Romulus and Remo.

correspond to an external node (in other words, a leaf of a real tree). Internal nodes represent a speciation event and the distance between two nodes is proportional to the divergence time.

In a phylogenetic tree derived from similarities between protein sequences, each external node is a protein and the distance between two nodes is inversely related to the similarity between the sequences of the two proteins. Later in this chapter we will describe an algorithm to build a phylogenetic tree from a set of sequences. Once

FIGURE 3.16 Phylogenetic tree of some mammals.

the tree has been built, multiple alignment methods will align sequences in adjacent nodes progressively until all sequences are aligned.

For example, given the tree in Figure 3.17, we will align first the sequences indicated by 2 and 3 and those indicated by 4 and 5, then we will proceed by aligning the first alignment with the sequence indicated by 1 and, finally, this alignment with the alignment of sequences 4 and 5. The next section will explain how we can align a sequence to an alignment and an alignment to an alignment.

This method is, by and large, the one used by CLUSTALW, probably the most used program for multiple alignments. This program also features other characteristics. It weights the sequences to compensate for their different levels of similarity, uses different similarity matrices according to the distance between the pair of sequences to be aligned, and takes advantage of the already built portion of the alignment to modify gap penalties.

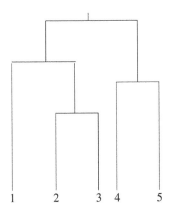

FIGURE 3.17 Example of a phylogenetic tree.

3.7.1 How to Add Sequences to a Pre-Existing Alignment

Now we need to explain how we can align a sequence to a pre-existing alignment. Once again, we will make use of a matrix. In the first column, we write the amino acids a_i of the new sequence and, in each cell of the first row, the amino acids A_{1j}, $A_{2j}, \ldots, A_{kj}, \ldots, A_{Nj}$ of the pre-existing alignment. The cell i,j will have a similarity value that is the average of the similarity values of residue a_i with $A_{1j}, A_{2j}, \ldots, A_{kj}, \ldots, A_{Nj}$.

Once the matrix is built, we can use the same method used to align two sequences to find the optimal alignment. For example, if we use PAM250, the matrix used to align the sequence AGRSGS to the alignment:

ASDKL

VSERF

is:

	A V	S S	D E	K R	L F
A	$\frac{1}{2}$ (PAM(A,A) + PAM(A,V)) = $\frac{1}{2}$ (2 + 0)	$\frac{1}{2}$ (1 + 1)	$\frac{1}{2}$ (0 + 0)	$\frac{1}{2}$ (−1 − 2)	$\frac{1}{2}$ (−2 − 4)
G	$\frac{1}{2}$ (1 − 1)	$\frac{1}{2}$ (1 + 1)	$\frac{1}{2}$ (1 + 0)	$\frac{1}{2}$ (−2 − 3)	$\frac{1}{2}$ (−4 − 5)
R	$\frac{1}{2}$ (−2 − 2)	$\frac{1}{2}$ (0 + 0)	$\frac{1}{2}$ (−1 − 1)	$\frac{1}{2}$ (3 + 6)	$\frac{1}{2}$ (−3 − 4)
S	$\frac{1}{2}$ (1 + 0)	$\frac{1}{2}$ (3 + 3)	$\frac{1}{2}$ (0 + 0)	$\frac{1}{2}$ (0 + 0)	$\frac{1}{2}$ (−3 − 3)
G	$\frac{1}{2}$ (1 − 1)	$\frac{1}{2}$ (1 + 1)	$\frac{1}{2}$ (1 + 0)	$\frac{1}{2}$ (−2 − 3)	$\frac{1}{2}$ (−4 − 5)
S	$\frac{1}{2}$ (1 + 0)	$\frac{1}{2}$ (3 + 3)	$\frac{1}{2}$ (0 + 0)	$\frac{1}{2}$ (0 + 0)	$\frac{1}{2}$ (−3 − 3)

We can also use hidden Markov models (HMMs) to align a sequence to a multiple alignment. The pre-existing multiple alignment is used to build an HMM, and the new sequence is aligned by calculating the path with the highest probability, similarly to what we discussed in the case of the splicing site method.

Generally, an HMM for a multiple alignment can be depicted as shown in Figure 3.18. In each position of the alignment, we can insert an amino acid, delete it, or align it to the column (we call the latter a "match state"). The probability of following one or the other path depends upon the transition probabilities that have been calculated from a pre-existing alignment. The alignment

AVFDFRT

AV-DYKT

AAFDSRT

will give rise to the HMM shown in Figure 3.18 where the arrows' thickness is proportional to the transitions probabilities.

We are not going to discuss it here, but it is worth mentioning that HMMs can also be used to build an alignment from individual sequences.

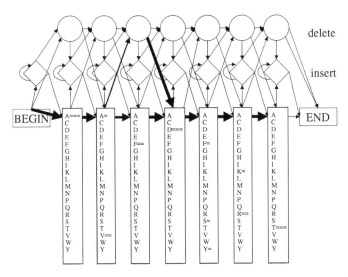

FIGURE 3.18 Hidden Markov model for a multiple sequence alignment of proteins.

3.8 PHYLOGENETIC TREES

We now know how to align two sequences, a sequence to an alignment, and two alignments; this is all we need to be able to build a phylogenetic tree that will allow us to visualize the relationship between sequences as a function of their evolutionary distance. The easiest definition of distance between two sequences is the percentage of nonidentical amino acids, but others can be devised.

Let us first consider each sequence separately and calculate its distance from all the others:

Percent different amino acids in the sequences	Seq 1	Seq 2	Seq 3	Seq 4
Seq 1	0	5	11	14
Seq 2		0	9	10
Seq 3			0	7
Seq 4				0

Next, we identify the pair of sequences with the lowest distance (if there is more than one pair, we choose one at random). In our example, the closest sequences are those indicated by 1 and 2 and their distance is 5. We will draw them, starting from an arbitrary origin, positioning the node between them at a height corresponding to half of their distance (in this way the sum of the heights of the two vertical segments will be equal to their distance) as shown in Figure 3.19.

We will call this set of two sequences cluster 1–2 and compute the distance between this cluster and each of the other sequences. The distance will be the average of the distances of each member of the cluster with each other sequence (for example,

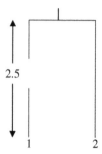

FIGURE 3.19 How to build a phylogenetic tree: step 1.

the distance between cluster 1–2 and sequence 3 is the average of the distances between sequence 1 and sequence 3 and the distance between sequence 2 and sequence 3):

	Cluster 1–2	3	4
Cluster 1–2	0	$\frac{1}{2}(d(1,3)+d(2,3))=10$	$\frac{1}{2}(d(1,4)+d(2,4))=12$
3		0	7
4			0

The lowest distance in the table is now the one between sequences 3 and 4, which can be included in our graph as shown in Figure 3.20. The same step will be repeated until we are left with only two clusters (Figure 3.21):

	Cluster 3–4
Cluster 1–2	$=\frac{1}{2}\,d(\text{cluster }12,3)+\frac{1}{2}\,d(\text{cluster }12,4)=11$

The tree-building method that we described is called *UPGMA*, which stands for unweighted pair group method using arithmetic averages. This suggests that there are methods which use different weights for the sequences and/or do not use the arithmetic averages to calculate the distances (for example, they use the minimum or maximum distance). Indeed, many tree-building algorithms are available, and UPGMA is neither the most used nor the most effective, but rather one of the most intuitive to understand. It is worth stressing that the algorithms to build phylogenetic trees heavily rely on the assumption that the alignments are correct.

How can we assess the quality of our tree? The most used method is called *bootstrap*. This method starts from the alignment used to build the tree, then builds many (e.g., 1,000) other alignments of the same size, choosing some columns of the original alignment randomly (the same column can be used more than once in the "artificial" alignments). A tree is built from each of the "artificial" alignments and, finally, for each branch of the tree, the number of times in which the same

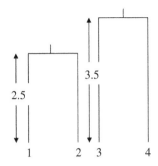

FIGURE 3.20 How to build a phylogenetic tree: step 2.

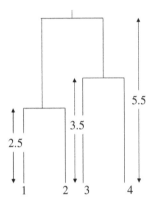

FIGURE 3.21 How to build a phylogenetic tree: step 3.

branch appears in the generated trees is counted. The branches present in more trees, independently on the columns used, are the most reliable.

REFERENCES

Historical Contributions

The historical paper of M. O. Dayhoff describing the PAM matrices has been published in Dayhoff, M. O., R. M. Schwartz, and B. C. Orcutt. 1978. A model for evolutionary change. In *Atlas of protein sequence and structure*, Washington, D.C.: National Biomedical Research Foundation, 5, 345–358.

The first alignment algorithm can be found in Needleman, S. B., and C. D. Wunsch. 1970. A general method applicable to the search for similarities in the amino acid sequence of two proteins. *Journal of Molecular Biology* 48:442–453.

The first method to calculate phylogenetic distances was suggested by Edwards, A. W. F., and L. Cavalli–Sforza. 1963. The reconstruction of evolution. *Annals of Human Genetics* 27:105–108.

Suggestions for Further Reading

The BLOSUM matrices are described in Henikoff, S., and J. G. Henikoff. 1992. Amino acid substitution matrices from protein blocks. *Proceedings of the National Academy of Science USA* 89:10915–10919.

CLUSTALW, the most used method for multiple sequence alignment, is described in Thompson, J. D., D. G. Higgins, and T. J. Gibson. 1994. Improved sensitivity of profile searches through the use of sequence weights and gap excision. *Computer Applications in the Biosciences* 10: 19–29.

An explanation on how to use HMM in multiple sequence alignment can be found in Karplus, K., C. Barrett, and R. Hughey. 1998. Hidden Markov models for detecting remote protein homologies. *Bioinformatics* 14:846–856.

A review of various methods to calculate phylogenetic distances can be found in Felsenstein, J. 1996. Inferring phylogenies from protein sequences by parsimony, distance and likelihood methods. *Methods in Enzymology* 266:418–427.

PROBLEMS

1. Find the EC classification of adenylsuccinate lyase and of arginilsuccinate lyase. How similar are they? Is their similarity reflected in their GO molecular function classification? Comment.

2. Align the following two sequences, using the PAM250 matrix and a –2 penalty for insertions and deletions. Report the final score:

A	C	F	A	T	F	A	R	T	G	H	I	L	T	G	G
L	C	W	W	V	G	S	T	S	G	H	L	A	P	W	F

3. Which regions of the full alignment that you can find in the entry PF04995.3 of the PFAM database would you use for deriving a substitution matrix? Compute the content of the cell corresponding to the A C pair of such a matrix.

4. How do you expect a PAM0 matrix to look? a PAM∞?

5. Compute the average score in the PAM250 matrix for hydrophobic–hydrophobic substitutions, hydrophobic–hydrophilic and hydrophilic–hydrophilic. Use the following amino acid class definition—hydrophobic: ACFGILPVW; hydrophilic: DEHKLMNQRST—and comment on the result.

6. Align the sequence ADFCTFSRTGSIVTAT to the alignment that you obtained in problem 2.

7. Build a phylogenetic tree, given the following table reporting the sequence identity between pairs of sequences:

	seq1	seq2	seq3	seq4	seq5	seq6
seq1	100					
seq2	65	100				
seq3	42	55	100			
seq4	70	35	40	100		
seq5	45	42	52	45	100	
seq6	40	28	25	15	30	100

8. Compute the transition probabilities for an HMM derived from the alignment obtained in problem 6. There are only three sequences in this alignment. Can you see why this can make the HMM less effective?

9. Which of the following matrices would you use to align sequences 3 and 5 of problem 7?
   PAM80
   BLOSUM80
   PAM200
   BLOSUM52

10. Retrieve the sequences contained in the full alignment of PFAM PF04995.3 and realign them using, for example, CLUSTAL. Comment on the differences, if any.

4 Similarity Searches in Databases

GLOSSARY

FASTA, BLAST: programs for similarity searching in databases

Low-complexity regions: regions of a sequence with a biased amino acid composition

PSI-BLAST: position-specific iterative BLAST; a version of BLAST in which a profile is automatically built from a multiple-sequence alignment of the significantly similar sequences found by an initial search with BLAST; the profile is used to perform a second search, whose results are used to refine the profile and so on, iteratively

Query sequence: sequence used as bait to search for similar sequences in a database

4.1 BASIC PRINCIPLES

In the previous chapter, we described some methods for aligning protein sequences and for evaluating the similarities between them. The next step is to devise a procedure to identify sequences homologous to the protein of interest among a set of protein sequences.

Two proteins are homologous if they have an evolutionary relationship—that is, if they derive from a common ancestor. Since, by and large, two homologous proteins are expected to share a higher sequence similarity than unrelated proteins, we can measure sequence similarity and use its value to estimate the probability that the two corresponding proteins are indeed homologous. In other words, we measure sequence similarity and, from it, we infer homology.

Clearly, in order to identify proteins homologous to our target (or *query*) protein, we need an algorithm to retrieve all proteins sharing a sufficiently high sequence similarity among those present in a database. How similar should two proteins be in order to be homologous (i.e., in order to allow us to infer that they originated from a common ancestral protein)? As usual, we rely on statistics: Two proteins are expected to be homologous if their sequences share a similarity higher than expected for two unrelated sequences.

Before tackling the issue of how we define "randomness" in this case, we need to ask ourselves a question. We have gene and protein sequence databases, so we can:

- search for similarities between the query protein sequence and the protein sequence databases
- search for similarities between the gene of the *query sequence* and the nucleotide sequence databases
- search for similarities between the query protein sequence and all the possible translations in amino acids (all six possible reading frames) of the nucleotide sequence databases

Which is best?

Wherever possible, it is certainly better to use the sequence of the gene product (the protein) rather than the gene sequence itself. First, DNA is composed of only four bases while proteins are composed of 20 amino acids; therefore, the probability of finding similarities by chance is higher for DNA sequences. Second, nucleotide databases are much more populated; hence, the background due to random similarities is higher. Next, for protein sequences, we can use similarity matrices reflecting their functional and evolutionary properties. Finally, evolution acts on function, which is brought about by the amino acids of the product, and therefore protein sequences are better conserved than gene sequences during evolution.

The methods described in this chapter apply to all three of the preceding cases, although we will mainly focus on the case of a similarity search between a query protein sequence and a protein sequence database.

4.2 THE METHODS

In a database search, we need to have a fast and efficient way to distinguish a similarity due to chance from a functionally and evolutionarily significant one. While it is easy to infer the presence of a true homology when two sequences are very similar, the identification of remote homologies (which are often the most biologically interesting) creates many more problems. If the evolutionary distance is very high, the expected sequence similarity can fall below the threshold expected for unrelated proteins, which are, after all, made by the same 20 amino acids.

As we mentioned, the identification of homologous proteins can be based on the calculation of the probability that the observed similarity is due to evolution rather than to chance alone. If proteins were random sequences of 20 amino acids, it would be rather easy to calculate the probability for two sequences, n and m amino acids long, to share a certain percentage of identical amino acids by chance. But protein sequences are not random and the 20 amino acids are not equally probable in protein sequences.

Our calculation of the probability that two sequences are homologous heavily depends upon our definition of "random" protein sequences. Up to one quarter of the amino acids of proteins belong to regions with repetitive sequences or to *low-complexity regions*. For example, a number of proteins contain a very high number of prolines, glutamines, or other amino acids (Figure 4.1). If our query protein contains such a low-complexity region and we do not take it into account in computing the probability of a randomly expected similarity score, all proteins with a

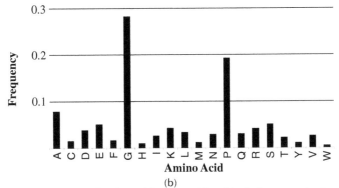

```
MMSFVQKGSW LLLALLHPTI ILAQQEAVEG GCSHLGQSYA DRDVWKPEPC QICVCDSGSV LCDDIICDDQ ELDCPNFEIP
FGECCAVCPQ PPTAPTRPPN GQGPQGPKGD PGPPGIPGRN GDPGIPGQPG SPGSPGPPGI CESCPTGPQN YSPQYDSYDV
KSGVAVGGLA GYPGPAGPPG PPGPPGTSGH PGSPGSPGYQ GPPGEPGQAG PSGPPGPPGA IGFSGPAGKD GESGRPGRPG
ERGLPGPPGI KGPAGIPGFP GMKGHRGFDG RNGEKGETGA PGLKGENGLP GENGAPGPMG PRGAPGERGR PGLPGAAGAR
GNDGARGSDG QPGPPGPPGT AGFPGSPGAK GEVGPAGSPG SNGAPGQRGE PGPQGHAGAQ GPPGPPGING SPGGKGEMGP
AGIPGAPGLM GARGPPGPAG ANGAPGLRGG AGEPGKNGAK GEPGPRGERG EAGIPGVPGA KGEDGKDGSF GEPGANGLPG
AAGERGAPGF RGPAGPNGIP GEKGPAGERG APGPAGPRGA AGEPGRDGVP GGPGMRGMPG SPGGPGSDGK PGPPGSQGES
GRPGPPGPSG PRGQPGVMGF PGPKGNDGAP GKNGERGGPG GPGPQGPPGK NGETGPQGPP GPTGPGGDKG DTGPPGPQGL
QGLPGTGGPP GENGKPGEPG PKGDAGAPGA PGGKGDAGAP GERGPPGLAG APGLRGGAGP PGPEGGKGAA GPPGFPGAAG
TPGLQGMPGE RGGLGSPGPK GDKGEPGGPG ADGVPGKDGP RGPTGPIGPP GPAGQPGDKG EGGAPGLPGI AGPRGSPGER
GETGPPGPAG FPGAPGQNGE PGGKGERGAP GEKGEGGPPG VAGPPGGSGP AGPPGPQGVK GERGSPGGPG AAGFPGAARGL
PGPPGSNGNP GPPGPSGSPG KDGPPGPAGN TGAPGSPGVS GPKGDAGQPG EKGSPGAQGP PGAPGPLGIA GITGARGLAG
PPGMPGPRGS PGPQGVKGES GKTGANGLSG ERGPPGPQGL PGLAGTAGEP GRDGNPGSDG LPGRDGSPGG KGDRGENGSP
GAPGAPGHPG PPGPVGPAGK SGDRGESGPA GPAGAPGPAG SRGAPGPQGP RGDKGETGER GAAGIKGHRG FPGNPGAPGS
PGPAGQQGAT GSPGPAGPRG PVGPSGPPGK DGTSGHPGPI GPPGPRGNRG RRGSRGSPGH PGQPGPPGPP GAPGPCGGGV
```

(a)

(b)

FIGURE 4.1 Sequence (a) and amino acids composition (b) of a low-complexity protein: the RT chain of human procollagen type III.

similarly biased composition will turn out to have a statistically significant score, independently of whether they share a common evolutionary origin with the query.

The available programs for database searching not only differ in the algorithm used, but also in their definition of random sequences. The most frequently used are *FASTA* and *BLAST* and their modifications. Different versions of these programs can differ slightly, but the rationale behind them does not and this is what we will describe next.

4.2.1 FASTA

FASTA divides the query sequence in "words"—that is, subsequences whose length can be chosen by the user (usually two or three amino acids). It then calculates the positions of the words in the query sequence and in all the sequences of the DB. The occurrence of each possible word in the database sequences is precomputed and stored to speed up the search. Let us assume that our database only contains the three sequences:

```
Seq 1:  LKDCDDAFSGSTLTLMRASRK
Seq 2:  ACKRAEFSGSVTRMLSTRK
Seq 3:  ACDDEFGLLLTRYTMASTRK
```

Also assume that we are using two-letters words. We can precalculate the position of the 400 possible pairs of amino acids in each of our database sequences:

Word	Seq 1	Seq 2	Seq 3
AC	—	1	1
CD	4	—	2
DD	5	—	3
DE	—	—	4
AE	—	5	—
EF	—	6	5
FG	—	—	6
GS	10	9	—
SA	—	—	—
AT	—	—	—
TR	—	12	11
RM	—	13	—
...	...	...	...
...	...	...	...
MA	—	—	15
AS	18	—	16
ST	11	16	17
RK	20	18	19

This table tells us that the amino acid pair CD is present in position 4 in the first sequence and in position 2 in the third sequence, while AE is present only in the second sequence in position 5. The same word can appear more than once in a given sequence without affecting the algorithm.

Given a query sequence—for example, ACDDEFGSATRMASTRK—we build a table containing the position of each amino acid pair in the query.

The next step is to calculate the offset—that is, the difference between the position of the word in the query and in each of the database sequences (Table 4.1). If more words have the same offset, they can be part of the same alignment without insertions or deletions. For example, the words GS, TR, RM, ST, and RK all share an offset of 2 with sequence 2 in the database (as shown in the sixth column of Table 4.1) and can be part of the alignment shown in Figure 4.2.

The next step in the FASTA algorithm is to calculate the score of each aligned region, using the PAM250 matrix, and to select the 10 highest scoring regions for each database sequence. The sum of the scores of these 10 regions (computed for each database sequence) is called init1. The database sequences are ordered according to init1 and the best N analyzed further. N depends upon the specific implementation of the program, the size of the database, etc.

Some aligned regions could be part of the same alignment if gaps are allowed (i.e., we could build an alignment including words with different offsets). For example, the words AC and EF have offsets 0 and 1, respectively, and can be included in the alignment containing the ones with offset 2 by allowing two gaps.

Query	A	C	—	D	D	E	F	—	G	S	A	T	R	M	A	S	T	R	K	
Position	1	2		3	4	5	6		7	8	9	10	11	12	13	14	15	16	17	
	X	X			X	X				X	X	X			X	X	X	X		
Seq 2	A	C	K	R	A	E	F	S	G	S	V	T	R	M	L	S	T	R	K	T
Position	1	2	3	4	5	6	7	8	9	10	11	12	13	14	15	16	17	18	19	20
Offset	0			1					2		2	2			2	2	2			

TABLE 4.1
Example of a FASTA Table for a Hypothetical Sequence and a Three-Sequence Database

Word	Query	Seq 1	Seq 2	Seq 3	Off 1 (Seq1—Query)	Off 2 (Seq2—Query)	Off 3 (Seq3—Query)
AC	1	—	1	1	—	0	0
CD	2	4	—	2	2	—	0
DD	3	5	—	3	2	—	0
DE	4	—	—	4	—	—	0
EF	5	—	6	5	—	1	0
FG	6	—	—	6	—	—	0
GS	7	10	9	—	3	2[a]	—
SA	8	—	—	—	—	—	—
AT	9	—	—	—	—	—	—
TR	10	—	12	11	—	2[a]	1
RM	11	—	13	—	—	2[a]	—
...	...	...	...	...	...	...	...
...	...	...	...	...	...	...	...
MA	12	—	—	15	—	—	3
AS	13	18	—	16	5	—	3
ST	14	11	16	17	-3	2[a]	3
RK	16	20	18	19	4	2[a]	3

[a] Shared offset with sequence 2 in the database.

Query	A	C	D	D	E	F	G	S	A	T	R	M	A	S	T	R	K	
Pos.	1	2	3	4	5	6	7	8	9	10	11	12	13	14	15	16	17	
					X	X			X	X			X	X	X	X		
Seq 2	A	C	K	R	A	E	F	S	G	S	V	T	R	M	L	S	T	R	K	T
Pos.	1	2	3	4	5	6	7	8	9	10	11	12	13	14	15	16	17	18	19	20
Offset						2			2	2			2	2						

FIGURE 4.2 Alignment on the basis of the data in Table 4.1.

The sum of the scores of these aligned regions, after subtracting a penalty accounting for the gaps, is the score initN.

Now the program reorders the sequences according to their initN score and performs a full-fledged sequence alignment of the query sequence with the best ones. The final score (called "opt") is calculated after this step and is therefore a proper alignment score. However, the reader should be aware that only sequences with a sufficiently high init1 and initN score are considered in this step.

Now we need to evaluate whether the final opt scores are statistically significant. Here is how FASTA tackles the problem. It generates a large number of random sequences with the same composition and length as those of the query sequence (to take into account possible unusual amino acid distributions). Next, it searches a subset of the database as described before using each of the random sequences as query and records all the scores. The result is the expected distribution of scores for random sequences. FASTA calculates the mean (M_{random}) and standard deviation (σ_{random}) of the distribution of the scores.

The opt score of the query is compared to this random distribution and the respective Z-score and the E-score are computed. The Z-score is the number of standard deviations separating the opt_{query} from the mean of the random distribution:

$$Z\text{-score} = (opt_{query} - M_{random})/\sigma_{random}$$

A high value (at least four) implies that the final score is unlikely to belong to the random distribution and therefore that the observed similarity is likely to be biologically significant.

The probability P that a certain alignment has a score equal or better than a value S follows a Poisson distribution:

$$P = 1 - e^{-K \cdot m \cdot n \cdot e^{-\lambda S}}$$

where m and n are the length of the two sequences. K and λ are parameters depending on the database and the matrix used and are estimated from the distribution of the scores in random sequence searches.

The value $E(S)$ is the number of sequences expected to have by chance the score S and can be shown to be:

$$E(S) = Kmn \cdot e^{-\lambda S}$$

The "bit score" S is defined as:

$$S' = \frac{\lambda S - \ln K}{\ln 2}$$

Therefore, if we use the bit score, we obtain:

$$E = mn2^{-S'}$$

The bit score takes into account the λ and K parameters. An increase of 1 in the bit score corresponds to a decrease of a factor $2^S = 2^1 = 2$ in the E-value, an increase of 10 corresponds to a decrease of a factor 1024 (2^{10}) in the E-value, and so on. This makes the bit score very useful for comparing results obtained in different database searches.

Figure 4.3(a,b) shows a typical output of a database search performed using FASTA.

4.2.2 BLAST

The original version of BLAST looks for contiguous similarity regions between the query and the database sequences (i.e., it does not take gaps into account). It is based on the idea that really similar proteins should have at least one uninterrupted region of similarity. A version of BLAST including gaps (called gapped BLAST) is also available and commonly used.

Similarly to FASTA, BLAST lists the words of a query sequence (generally three amino acids long) and compares them with regions of equal length in the database sequences. The score of each word (using, for example, the BLOSUM62 matrix) is compared to a given threshold T. Only words with a score greater than or equal to T are further analyzed; that is, only words containing amino acids that have on average a high matching score are considered. For each of them, BLAST analyzes the matching sequences in the database and tries to extend the alignment in both directions, until the score falls below another threshold S (Figure 4.4).

The main difference between FASTA and BLAST, which also accounts for the better time performance of the latter, depends upon the way the two programs compute the distribution of randomly expected scores. While FASTA calculates the expected distribution for each query sequence, BLAST assumes that the query sequence has an average composition. It therefore compares the scores of the query sequence with a precomputed distribution that only depends upon the composition of the database and the matrix used. In other words, similarly to FASTA, BLAST calculates $E(S)$ as:

$$E = Kmn \cdot e^{-\lambda S}$$

where m is the query sequence length, as in FASTA, but n is the total length of the sequences in the database and therefore is always the same for each search in the same database. K and λ are also calculated a priori, as we mentioned. BLAST uses a set of random sequences with a standard composition (therefore, it does not need to repeat the calculation for each query).

Of course, this could lead to problems for sequences with unusual composition. BLAST deals with this problem by masking regions with "unusual" compositions (the program replaces them with the symbol "X") and excludes them from the search.

However, other properties of regions of proteins not directly related to their amino acid sequence, such as the presence of coiled-coil or transmembrane helices, might affect the statistics of the score distribution. There is the risk that the scores reported for these regions with nonhomologous proteins sharing the same properties are higher than expected for a sequence with a standard composition. If we are

```
FASTA searches a protein or DNA sequence data bank
 version 3.3t09 May 18, 2001
Please cite:
 W.R. Pearson & D.J. Lipman PNAS (1988) 85:2444-2448

@:1-: 138 aa
 EMBOSS_001
 vs  SWISS-PROT All library
searching /ebi/services/idata/fastadb/swall library

157883612 residues in 463388 sequences
 statistics extrapolated from 60000 to 463161 sequences
  Expectation_n fit: rho(ln(x))= 5.6891+/-0.000178; mu= 0.2866+/- 0.010
 mean_var=73.7525+/-14.770, 0's: 157 Z-trim: 35  B-trim: 0 in 0/65
 Lambda= 0.1493

FASTA (3.39 May 2001) function [optimized, BL50 matrix (15:-5)] ktup: 2
 join: 36, opt: 24, gap-pen: -12/ -2, width:  16
 Scan time:  7.890

The best scores are:                                       opt bits   E
SWALL:FLAV_CLOBE P00322 FLAVODOXIN.             ( 138)  915  206   8e-53
SWALL:FLAV_MEGEL P00321 FLAVODOXIN.             ( 137)  435  103  1.1e-21
SWALL:FLAV_TREPA O83895 FLAVODOXIN.             ( 146)  355   85  1.8e-16
SWALL:FLAV_DESSA P18086 FLAVODOXIN.             ( 146)  234   59  1.2e-08
SWALL:FLAV_DESGI Q01095 FLAVODOXIN.             ( 146)  230   59  2.3e-08
SWALL:FLAW_DESGI Q01096 FLAVODOXIN.             ( 147)  225   57  4.8e-08

                        ...omitted...

SWALL:FLAV_SYNP7 P10340 FLAVODOXIN.             ( 169)  124   36   0.19
SWALL:FLAV_ENTAG P28579 FLAVODOXIN.             ( 177)  124   36    0.2
SWALL:Q9JVW2 Q9JVW2 PUTATIVE PILUS ASSEMBLY PROTE ( 371)  128   37   0.21
SWALL:Q9JY02 Q9JY02 PILM PROTEIN.               ( 371)  128   37   0.21
SWALL:Q9PMR8 Q9PMR8 FLAVODOXIN.                 ( 163)  123   37   0.22
SWALL:Q50531 Q50531 FLAVOPROTEIN.               ( 404)  128   37   0.22
SWALL:O27404 O27404 FLAVOPROTEIN AI.            ( 409)  128   37   0.22
SWALL:O26322 O26322 FLAVOPROTEIN A HOMOLOG (II). ( 410)  128   37   0.22
SWALL:O05130 O05130 PILM.                       ( 371)  126   36   0.28
SWALL:Q9CNS2 Q9CNS2 FLDA.                        ( 174)  121   35   0.31
SWALL:ROO_DESGI Q9F0J6 RUBREDOXIN-OXYGEN OXIDORED ( 402)  123   36   0.47
SWALL:O30070 O30070 FLAVOPROTEIN (FPRA-1).      ( 336)  121   35   0.54

>>SWALL:FLAV_CLOBE P00322 FLAVODOXIN.                    (138 aa)
 initn: 915 init1: 915 opt: 915  Z-score: 1082.5  bits: 206.1 E(): 8e-53
Smith-Waterman score: 915;  100.000% identity (100.000% ungapped) in 138 aa overlap (1-138:1-138)

              10        20        30        40        50        60
EMBOSS MKIVYWSGTGNTEKMAELIAKGIIESGKDVNTINVSDVNIDELLNEDILILGCSAMGDEV
       ::::::::::::::::::::::::::::::::::::::::::::::::::::::::::::::::
SWALL: MKIVYWSGTGNTEKMAELIAKGIIESGKDVNTINVSDVNIDELLNEDILILGCSAMGDEV
              10        20        30        40        50        60

              70        80        90       100       110       120
EMBOSS LEESEFEPFIEEISTKISGKKVALFGSYGWGDGKWMRDFEERMNGYGCVVVETPLIVQNE
       ::::::::::::::::::::::::::::::::::::::::::::::::::::::::::::::::
SWALL: LEESEFEPFIEEISTKISGKKVALFGSYGWGDGKWMRDFEERMNGYGCVVVETPLIVQNE
              70        80        90       100       110       120

             130
EMBOSS PDEAEQDCIEFGKKIANI
       ::::::::::::::::::::
SWALL: PDEAEQDCIEFGKKIANI
             130

>>SWALL:FLAV_MEGEL P00321 FLAVODOXIN.                    (137 aa)
 initn: 376 init1: 376 opt: 435  Z-score: 523.6  bits: 102.7 E(): 1.1e-21
Smith-Waterman score: 435;  46.324% identity (47.015% ungapped) in 136 aa overlap (1-136:2-135)

              10        20        30        40        50        60
EMBOSS  MKIVYWSGTGNTEKMAELIAKGIIESGKDVNTINVSDVNIDELLNEDILILGCSAMGDE
        ..:::::::::: ::. :  ..  .: ::...    :..: :....::::::..:::.:
SWALL: MVEIVYWSGTGNTEAMANEIEAAVKAAGADVESVRFEDTNVDDVASKDVILLGCPAMGSE
              10        20        30        40        50        60

       60        70        80        90       100       110
EMBOSS VLEESEFEPFIEEISTKISGKKVALFGSYGWGDGKWMRDFEERMNGYGCVVVETPLIVQN
```

(a)

FIGURE 4.3 Example of a FASTA result for a protein sequences database search.

```
            ::.:  :::. ... :..::::.::::::::::.:.:: ...: . : .:. : ::..
  SWALL: ELEDSVVEPFFTDLAPKLKGKKVGLFGSYGWGSGEWMDAWKQRTEDTGATVIGTA-IVNE
              70        80        90       100       110

     120       130
  EMBOSS EPDEAEQDCIEFGKKIANI
         ::.: . : :.:. :.
  SWALL: MPDNAPE-CKELGEAAAKA
         120       130

  >>SWALL:FLAV_TREPA O83895 FLAVODOXIN.                    (146 aa)
   initn: 286 init1: 263 opt: 355  Z-score: 430.0  bits: 85.4 E(): 1.8e-16
  Smith-Waterman score: 355;  42.143% identity (44.030% ungapped) in 140 aa overlap (3-136:6-145)

              10        20        30        40        50
  EMBOSS    MKIVYWSGTGNTEKMAELIAKGIIESGKDVNTINVSDVNIDELLNEDILILGCSAMG
            ...:::::.:: ::. :..:. .: ..::: :.. . . : ..:::: :
  SWALL: MAKVAVIFWSGTGHTETMARCIVEGLNVGGAKADLFSVMDFDVGTFDSYDRFAFGCSAAG
            10        20        30        40        50        60

        60        70        80        90       100       110
  EMBOSS DEVLEESEFEPFIEEISTKISGKKVALFGSYGW-GDGK---WMRDFEERMNGYGCVVVE-
         .: :: ::::::. : ..::::::::::: : :.:.  ::.. :: .. : :
  SWALL: SEELESSEFEPFFTSIEGRLSGKKVALFGSYEWAGEGEGGEWMVNWVERCKAAGADVFEG
              70        80        90       100       110       120

       120       130
                                (b)
```

FIGURE 4.3 (Continued)

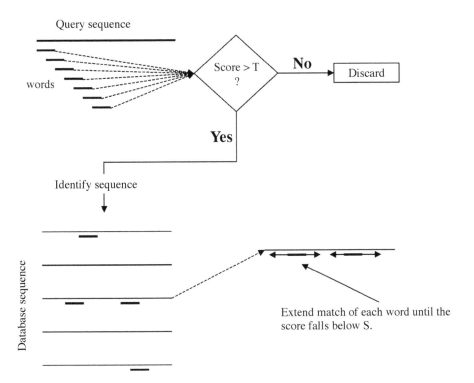

FIGURE 4.4 Schematic representation of the algorithm used by BLAST.

interested in the results of a limited number of database searches—that is, we are not, for example, analyzing all the proteins of a large genome—it is highly advisable to look at the alignment and to use our biological knowledge of the query protein to decide whether a similarity is biologically relevant. In other cases, we can compare the results obtained using different programs and masking algorithms.

Recently, BLAST has implemented two methods for taking biased composition into account. The first method adjusts all the values of the amino acid substitution matrices by an analytically determined constant. The second, which is more sophisticated and slow, is to adjust each score separately to compensate for the compositions of the two sequences being compared.

An example of the output of BLAST is shown in Figure 4.5(a,b).

4.2.3 PROFILE SEARCHES

It has already been mentioned how useful a multiple alignment can be for unraveling the relationships between proteins, and we have also discussed how we can align a sequence to a multiple alignment. If we now focus our attention on how the alignment is scored, we can notice another advantage of multiple alignments, or profiles or HMM: They automatically assign a relatively higher score to more conserved regions.

Let us go back to the example used in the previous chapter; the similarity matrix between the sequence AGRSGS and the sequence ASDKL is:

	A	S	D	K	L
A	2	1	0	−1	−2
G	1	1	1	−2	−4
R	−2	0	−1	3	−3
S	1	3	0	0	−3
G	1	1	1	−2	−4
S	1	3	0	0	−3

Let us compare it with the similarity matrix between the sequence AGRSGS and the alignment:

ASDKL

VSERF

	A V	S S	D E	K R	L F
A	1	1	0	−1.5	−3
G	0	1	0.5	−2.5	−4.5
R	−2	0	−1	4.5	−3.5
S	0.5	3	0	0	−3
G	0	1	0.5	−2.5	−4.5
S	0.5	3	0	0	3

Comparing the two matrices, it becomes clear that matches involving amino acids conserved in the alignment will receive a comparatively higher score if they are also conserved in the final alignment. This implies that it is more convenient to use multiple alignments as query in our database search. Often, the multiple sequence alignment is encoded in the form of a profile (i.e., a position-specific matrix). Also, HMMs can be used as query in a database search. The score of each sequence of the database will be the probability value associated with its optimal path.

The other advantage in using multiple alignments as query is that, by using the information contained in the protein profile, we can have a better idea of the likely position of gaps.

4.2.4 PSI-BLAST

Even if we do not have a multiple-sequence alignment of our protein of interest with other related proteins, an initial database search will almost certainly provide us with a list of proteins likely to be homologous to our query. These sequences can be used to build a multiple-sequence alignment that we can use as query for subsequent searches.

This procedure can be easily automated and it is indeed the basis of the *PSI-BLAST* (position-specific iterated BLAST) program, which works according to the following procedure:

1. Given a query sequence, PSI-BLAST uses gapped BLAST to perform a database search.
2. The program builds a multiple alignment including all sequences sharing a statistically significant score with the query.
3. This multiple alignment is used to build a profile, which is used as input for a new round of database searches.
4. PSI-BLAST estimates the statistical significance of the scores obtained in this new search and adds the sequences with a significant score to the multiple alignment.
5. It goes back to step 3 until no new sequences with a significant score are found or until it reaches a user-defined number of steps.

The method used by BLAST to evaluate the probability that a given score is statistically significant has been discussed before. But how can we decide whether a score is significant when it is obtained by aligning a sequence to a profile? This is clearly a crucial step if we want to run the procedure in a completely automated fashion.

Unfortunately, the statistics in this case is very complex and there is no simple analytical solution. BLAST resorts to applying the same statistics used for pair-wise alignments with the same values of K and λ.

Despite the necessary approximations, the results of PSI-BLAST are quite reliable and many interesting evolutionary relationships have been discovered thanks to this program. It is, however, a good idea not to wait until the program reaches convergence (i.e., stopping the procedure only when no new sequences are found),

```
Reference:
Altschul, Stephen F., Thomas L. Madden, Alejandro A. Schäffer,
Jinghui Zhang, Zheng Zhang, Webb Miller, and David J. Lipman (1997),
"Gapped BLAST and PSI-BLAST: a new generation of protein database search
programs", Nucleic Acids Res. 25:3389-3402.

RID: 999098351-6668-24212

Query=
        (138 letters)

Database: nr
           751,767 sequences; 239,157,582 total letters

                                                        Score     E
Sequences producing significant alignments:            (bits)  Value

gi|120274|sp|P00322|FLAV_CLOBE  FLAVODOXIN >gi|65882|pir||FX...   208   5e-54
gi|15023456|gb|AAK78565.1|AE007574_3  (AE007574) Flavodoxin ...   112   7e-25
gi|15011963|gb|AAK77653.1|  (AF109075) FlxD [Clostridium dif...   101   1e-21
gi|120281|sp|P00321|FLAV_MEGEL  FLAVODOXIN >gi|65881|pir||FX...    91   2e-18
gi|229364|prf||711671A  flavodoxin [Megasphaera elsdenii]         74   2e-13
gi|6015165|sp|O83895|FLAV_TREPA  FLAVODOXIN >gi|7430914|pir|...    63   5e-10
gi|15025470|gb|AAK80406.1|AE007745_5  (AE007745) Flavodoxin ...    50   3e-06
gi|13786656|pdb|1F4P|A  Chain A, Y98w Flavodoxin Mutant 1.5a...    49   8e-06

                              ...omitted...

gi|14972783|gb|AAK75401.1|  (AE007429) flavodoxin [Streptoco...    31   1.5
gi|13897519|gb|AAK48421.1|AF208233_1  (AF208233) pyruvate:fe...    30   2.6
gi|1168402|sp|P42058|ALA7_ALTAL  MINOR ALLERGEN ALT A 7 (ALT...    30   2.7
gi|9937183|gb|AAG02328.1|  (AF190966) envelope glycoprotein ...    30   3.0
gi|15283886|ref|NP_203043.1|  flavodoxin [Aquifex aeolicus] ...    30   4.2
gi|11362782|pir||D82719  tryptophan repressor binding protei...    30   4.4
gi|12723505|gb|AAK04706.1|AE006294_1  (AE006294) flavodoxin ...    29   5.1
gi|168870|gb|AAA33612.1|  (M13208)  qa-1S [Neurospora crassa]     29   6.9
gi|131758|sp|P11637|QA1S_NEUCR  QUINATE REPRESSOR >gi|83793|...    29   6.9

                              Alignments
Score = 208 bits (530), Expect = 5e-54
 Identities = 112/138 (81%), Positives = 112/138 (81%)

Query: 1    MKIVYWSGTGNTEKMAELIAKGIIESGKDVNTINVSDVXXXXXXXXXXXXXXGCSAMGDEV 60
            MKIVYWSGTGNTEKMAELIAKGIIESGKDVNTINVSDV            GCSAMGDEV
Sbjct: 1    MKIVYWSGTGNTEKMAELIAKGIIESGKDVNTINVSDVNIDELLNEDILILGCSAMGDEV 60

Query: 61   LXXXXXXXXXXXXXXTKISGKKVALFGSYGWGDGKWMRDFEERMNGYGCVVVETPLIVQNE 120
            L             TKISGKKVALFGSYGWGDGKWMRDFEERMNGYGCVVVETPLIVQNE
Sbjct: 61   LEESEFEPFIEEISTKISGKKVALFGSYGWGDGKWMRDFEERMNGYGCVVVETPLIVQNE 120

Query: 121  PDEAEQDCIEFGKKIANI 138
            PDEAEQDCIEFGKKIANI
Sbjct: 121  PDEAEQDCIEFGKKIANI 138

>gi|1941940|pdb|3NLL|   Clostridium Beijerinckii Flavodoxin Mutant: G57a Oxidized
          Length = 138

 Score =  112 bits (279), Expect = 7e-25
 Identities = 67/138 (48%), Positives = 82/138 (58%), Gaps = 1/138 (0%)

Query: 1    MKIVYWSGTGNTEKMAELIAKGIIESGKDVNTINVSDVXXXXXXXXXXXXXXGCSAMGDEV 60
            + I+Y SGTGNTE MA LI++G  +G DV  INVSD          G  AMGDEV
Sbjct: 4    INIIYCSGTGNTEAMANLISEGAKSAGADVQLINVSDADVDKVKNADVIVLGSPAMGDEV 63

Query: 61   LXXXXXXXXXXXXXXTKISGKKVALFGSYGWGDGKWMRDFEERMNGYGCVVVETPLIVQNE 120
            L            ++ GKKVALFGSYGWGDG++MRD+ ERM GYG  ++  LIVQ+
Sbjct: 64   LEEEEMEPFVENISKEVKGKKVALFGSYGWGDGQFMRDWVERMEGYGADLIGEGLIVQDA 123

Query: 121  PD-EAEQDCIEFGKKIAN 137
            P+ E E C EFGK + N
Sbjct: 124  PEGETEDQCREFGKALIN 141

>gi|15011963|gb|AAK77653.1|  (AF109075) FlxD [Clostridium difficile]
          Length = 142

 Score =  101 bits (251), Expect = 1e-21
```

(a)

FIGURE 4.5 Example of the results of a database search using BLAST. Note the use of the "X" character to mask the region 39–57 of the query. (The sequence is NIDELLNEDILIL.)

```
Identities = 55/136 (40%), Positives = 78/136 (56%), Gaps = 2/136 (1%)

Query:   3   IVYWSGTGNTEKMAELIAKGIIESGKDVNTINVSDVXXXXXXXXXXXXXXXGCSAMGDEVLX  62
             IVYWSGTGNTEKMA  +A+G+   GK   ++VS +           GC +MG E L
Sbjct:   6   IVYWSGTGNTEKMANFVAEGVKLKGKTPEVLDVSLLKPSDLKEEDKFALGCPSMGAEQLE  65

Query:  63   XXXXXXXXXXXXXXTKISGKKVALFGSYGWGDGKWMRDFEERMNGYGCVVV--ETPLIVQNE  120
                 + +SGK++ LFGSYGWG+ +WMRD+EERM   G  ++   E     +++
Sbjct:  66   EGDMEPFVSELESMVSGKQIGLFGSYGWGNCEWMRDWEERMQNAGATIIGGEGITAMEDP  125

Query: 121   PDEAEQDCIEFGKKIA  136
                +EA+ +CIE GK +A
Sbjct: 126   NEEAKDECIELGKTLA  141

>gi|120281|sp|P00321|FLAV_MEGEL FLAVODOXIN
 gi|65881|pir||FXME flavodoxin - Megasphaera elsdenii
         Length = 137

 Score = 90.5 bits (223), Expect = 2e-18
 Identities = 54/136 (39%), Positives = 68/136 (49%), Gaps = 2/136 (1%)

Query:   1   MKIVYWSGTGNTEKMAELIAKGIIESGKDVNTINVSDVXXXXXXXXXXXXXXXGCSAMGDEV  60
             ++IVYWSGTGNTE MA  I  +  +G DV ++   D        GC AMG E
Sbjct:   2   VEIVYWSGTGNTEAMANEIEAAVKAAGADVESVRFEDTNVDDVASKDVILLGCPAMGSEE  61

Query:  61   LXXXXXXXXXXXXXXXTKISGKKVALFGSYGWGDGKWMRDFEERMNGYGCVVVETPLIVQNE  120
             L                K+ GKKV LFGSYGWG G+WM +++R   G  V+ T  IV
```

(b)

FIGURE 4.5 (Continued.)

but rather to set a limit to the number of cycles (generally between three and five) and inspect the output of the program before deciding whether or not it is appropriate to continue with the iterative procedure.

REFERENCES

Historical Contributions

The first algorithm for database searches: FASTA:

Lipman, D. J., and Pearson, W. R. 1985. Rapid and sensitive protein similarity searches. *Science* 227:1435–1441.

The description of BLAST and PSI-BLAST:

Altschul, S. F., W. Gish, et al. 1990. Basic local alignment search tool. *Journal of Molecular Biology* 215:403–410.

Altschul, S. F., and E. V. Koonin. 1998. Iterated profile searches with PSI-BLAST—A tool for discovery in protein databases. *Trends in Biochemical Sciences* 23:444–447.

Suggestions for Further Reading

A comprehensive description of the methods to compare sequences can be found in Pearson, W. R. 1996. Effective protein sequence comparison. *Methods in Enzymology* 266:227–258.

PROBLEMS

1. Use the region spanning amino acids 300–482 of histone deacetylase 1 (SwissProt code: HDAC1_HUMAN) as query sequence for a Blast search on the nr database in three different modes: with low-complexity filter; with composition-based statistics and no masking; and with no masking and no adjustment. Compare the 10 highest scoring sequences in the three cases and comment on the results. Which of the three searches do you think provides more information and why?

2. Derive a profile for the following alignment. Transform it in the form of a log-odd table.

```
QVCGRGYARAWI.EVCG..ASV....GR.LA.
KACGRELVRLWV.EIC...GSV....SW.GR.
KLCGRELVRAQI.AIC...GMSTW.SKRSLSQ
HLCGSHLVEALYILVC...GE......R.GFF
HLCGPHLVEALY.LVC...GE......R.GFF
HLCGSHLVEALY.LVC...GE......R.GFF
HLCGSHLVEALY.LVC...GE......R.GFF
HLCGPHLVEALY.LVC...GE......R.GFF
HLCGSNLVEALY.MTC...GR......S.GF.
HLCGSNLVETLY.SVC...QD......D.GFF
HLCGSHLVEALY.LVC...GE......R.GFF
HLCGSHLVEALY.LVC...GD......R.GFF
HLCGSHLVDALY.LVC...GE......K.GFF
HLCGSHLVDALY.LVC...GD......R.GFF
HLCGSHLVDALY.LVC...GP......T.GFF
HLCGKDLVNALY.IAC...GV......R.GFF
YLCGSTLADVLS.FVC...GN......R.GY.
TLCGSELVDTLQ.FVC...DD......R.GFF
TLCGGELVDALQ.FVC...ED......R.GFY
TLCGGELVDTLQ.FVC...GD......R.GFY
TLCGAELVDALQ.FVC...GD......R.GFY
IYCGRYLAYKMA.DLCW..RAGF..EKRSVAH
TYCGRHLARTLA.NLCWEAGVD....KRSDAQ
TYCGRHLARTMA.DLCW..EEGV..DKRSDAQ
FYCGDFLARTMS.SLCW..SDMQ...KRSGSQ
TYCGRHLANILA.YVCF..GVE....KRGGAQ
TYCGRHLADTLA.DLCF..GVE....KRGGAQ
RYCGRVLADTLA.YLC...PEMEEVEKRSGAQ
```

3. Compute the score of the alignment between the following sequence and the profile derived in problem 2: QVCGRGYARAWILEVCGGAAS-VIGDEGRILA.

4. Repeat the search of problem 1 using tblastN. Explain what this means and comment on the results.

5. Here is the hypothetical result of a database search. Let us assume that we know that our query protein is a growth factor. Compute an ROC for the database search method for different values of an E-value threshold.

Hit	Score	E-value	Description
1	482	1E-135	Vascular endothelial growth factor isoform g precursor (*Homo sapiens*)
2	462	2E-129	Vascular endothelial growth factor isoform f precursor (*Homo sapiens*)
3	383	9E-106	Vascular endothelial growth factor isoform e precursor (*Homo sapiens*)
4	380	6E-105	Vascular endothelial growth factor isoform d precursor (*Homo sapiens*)
5	373	7E-103	Neurotrophin-7; growth factor; NT-7 (*Danio rerio*)
6	372	2E-102	Myotrophin (*Rattus norvegicus*)
7	367	7E-101	Growth factor (vaccinia virus)
8	313	1E-84	Growth factor (vaccinia virus)
9	312	2E-84	Growth factor (vaccinia virus)
10	295	3E-79	Growth factor (*Mustela vison*)
11	286	1E-76	Growth factor (lumpy skin disease virus)
12	252	3E-66	Serine/threonine-protein kinase
13	246	2E-64	Growth factor (*Griffithsia japonica*)
14	245	2E-64	Growth factor (*Escherichia coli*)
15	242	3E-63	Growth factor (cowpox virus)
16	236	1E-61	Growth factor
17	236	2E-61	Growth factor
18	233	9E-61	Growth factor
19	232	2E-60	Growth factor
20	222	2E-57	Growth factor
21	204	6E-52	Vascular endothelial growth factor isoform c precursor (*Homo sapiens*)
22	149	3E-35	Alpha-glucosidase precursor
23	58	4E-08	Serine/threonine-protein kinase
24	57	9E-08	Transforming growth factor, beta 1 (*Mus musculus*)
25	57	1E-07	Proteoglycan-4 precursor
26	55	3E-07	P40 T-cell and mast cell growth factor precursor
27	51	0.000007	Protein tonB
28	50	0.00001	Huntingtin
29	49	0.00003	Hepatocyte growth factor isoform 5 precursor (*Homo sapiens*)
30	49	0.00003	Bactenecin-7 precursor
31	37	0.097	61-kDa protein homologue
32	37	0.097	Transcription factor RelB
33	37	0.13	Homeobox protein Nkx-2.3
34	35	0.63	Large proline-rich protein BAT2
35	35	0.63	Cyclin-dependent kinase
36	34	0.82	Histone deacetylase 9
37	34	0.82	Putative transcription factor

6. Which *E*-value would you select?
7. Compute sensitivity and specificity of the database search method used in problem 5 when an E-value threshold of 10^{-5} is used.
8. Find (in SwissProt) the FHIT protein (accession P49789), a protein related to the human fragile histidine triad protein. Is it related to human galactose-1-phosphate uridylyltransferases?
9. How would you use Blast to verify whether the following pair of primers is appropriate to clone a human gene? Which gene would they detect?
   primer 1: GTCAAGTGGCAACTCCGTCAG
   primer 2: TTGAGAGATGGATTGTTGCGC
10. Discuss the problem related to multidomain proteins in database searches. (Notice the parameters used for computing the probability of the significance of a match.)

5 Amino Acid Sequence Analysis

GLOSSARY

Helical wheel: projection on a plane of a sequence assuming it is folded as an α-helix

Hydrophobicity plot: plot of the average value of amino acids' hydrophobicity calculated in a window running along a sequence

Orphan sequence: a sequence with no clear evolutionary relationship with any known sequence

5.1 BASIC PRINCIPLES

Sometimes, admittedly increasingly rarely, a similarity search does not find any sequence significantly similar to our query. This is a nightmare for any bioinformatician! It is very frustrating to look at a long series of amino acids with no additional information, especially if a biologist is staring at you with a very hopeful and impatient look.

In this chapter we will describe some of the methods that can be applied in these cases. In the next chapter we will describe methods that can be used to infer the three-dimensional structure of a protein since, in a significant number of cases, the latter can be very helpful in understanding the biological properties of our protein.

Protein sequences with no apparent similarity to any other known protein are called *orphan sequences*. In some cases, they could even not be a protein. If they are the product of hypothetical genes, it is possible that they are indeed never translated into a product. Even if this is not the case, with only one single sequence available, there is no way to detect which amino acids have a functional or structural role in the protein.

However, a sequence might only be apparently an orphan; it could be homologous to other proteins but share with them a sequence similarity below the detection threshold of the database search methods described in the previous chapter. For example, this is the case of many proteins encoded by RNA-viruses (i.e., viruses that use RNA rather than DNA as genetic material). The fidelity of RNA replication is many-fold lower than that of DNA, and viral sequences can accumulate many mutations in a short time sometimes making the detection of evolutionary relationships almost impossible.

5.1.1 Is It Really an Orphan Sequence?

It should be clear by now that using the default parameters and programs for database searches is not necessarily sufficient in order to search (and, more importantly, find) sequences similar to our query sequence in a database. Database searching is a complex problem, and the proposed solutions are all based on approximations and parameterizations. Before accepting the fact that our sequence is indeed an orphan sequence, we need to make sure that this is the case, without leaving any possibility unattended. As we will discuss, if a sequence is really unique, there is not much that we can say about the encoded protein. Therefore, before looking at the methods available in this case, it is convenient to discuss some of the strategies that we can try to use to increase the sensitivity of our database searches.

For example, there exist derived data banks containing collections of multiple alignments. Very often a search in these databases is more informative than a simple search in a sequence database, since we can exploit all the power of the multiple sequence alignment methods discussed in the previous chapter. Failing this, we can also search our protein sequence in nucleotide and genomics databases. It is indeed possible that a protein similar to our query is present in one of the sequenced genomes, but has not been identified by gene-finding methods.

Finally, many proteins are made by domains (i.e., independently folding units that are sometimes endowed with a specific function). If a protein is large (>300 amino acids), the probability that it is made of different modules or domains is quite high. We should remember that the methods for evaluating the score of a match between two sequences depend on the length of the two sequences; therefore, a similarity limited to a region of the query could fall below the significance threshold expected for a match spanning the full length of the sequence. Some methods try to identify domain boundaries within protein sequences; some reach respectable accuracy in some cases. Even if they do not provide a useful suggestion about the size and location of putative domains, it is worth splitting the sequence in overlapping regions 100–200 amino acids long and searching the database with each of them.

We also need to keep in mind that database search programs take advantage of several approximations to speed up the procedure, among which is preselection of database sequences sharing at least one "word" with the query. If our database search does not report any significant match, it is worth repeating the search with one-letter words.

Other problems could arise from the statistical approximation used by BLAST of considering a standard composition for a sequence and from masking. Sometimes, important regions can be flagged as repetitive and not considered in the search. We can always recur to FASTA or use different masking methods or none altogether.

Finally, a certain method can sometimes be useful: a "property search." This means searching for sequences sharing specific properties with the query such as the approximate length, amino acid composition, isoelectric point, etc. and there are programs for doing this. All the retrieved sequences need to be subsequently aligned and it is even more important in this case to analyze the results very carefully.

As far as the evaluation of the results is concerned, it is also important to remember that each entry of a database also contains bibliographic references to the scientific articles describing what is known about each protein and that, after all, it is always possible to perform experiments.

5.2 SEARCH FOR SEQUENCE PATTERNS

Some sequence patterns and the databases storing them have already been mentioned. The presence of one or more of these patterns in an orphan sequence can help us in identifying, at least, its functional class. Let us illustrate this with an example.

Proteases are enzymes catalyzing the hydrolysis of a peptide bond. There are different types of proteases, with different mechanisms of action. A very populated class is represented by serine proteases, which are, in turn, subdivided in different subclasses, one being the chymotrypsin-like serine protease family. The catalytic mechanism involves three amino acids: an aspartic acid, a histidine, and a serine (from which the name of the class is derived).

These three amino acids are far in the amino acid sequence but come close to each other in space when the protein is correctly folded; they must interact among themselves and with the substrate in order for the bond of the substrate to be hydrolyzed. We can say that the remaining amino acids of the protein are there only to guarantee that these amino acids, forming the active site, are correctly placed in space. However, this gives us at least a starting point for analyzing the protein sequence.

First, serine proteases are a family of homologous proteins; hence, their structure is generally conserved. Therefore:

- The order in the sequence of the three amino acids of the active site is always the same: His, followed by Asp, followed by Ser.
- The carboxyl group of the substrate must be positioned in a polar pocket making specific hydrogen bonds. This is made possible by two glycine residues (or a glycine and a serine) next to the catalytic serine.

As always, we can recur to the multiple sequence alignments of proteins of the family to identify other common features:

- The catalytic His is almost always before a Cys and after a big hydrophobic residue, a small polar residue, an Ala, and a small amino acid.
- The catalytic Ser is followed by two Gly (or a Gly and an Ser), as already mentioned, and preceded by a negatively charged residue and another Gly.
- Other regions upstream from the catalytic Ser and His also share some conserved characteristic features.

Each of these features can be written as a sequence pattern, and they are indeed stored in the PROSITE database:

[LIVM]-[ST]-A-[STAG]-H-C
[DNSTAGC]-[GSTAPIMVQH]-x(2)-G-[DE]-S-G-[GS]-[SAPHV]- [LIVM-
FYWH]-[LIVMFYSTANQH]

The first pattern reads: one amino acid among Leu, Ile, Val, and Met, then an amino acid that is an Ser or a Thr, an alanine followed by a small amino acid (Ser, Thr, Ala, or Gly), a histidine, and a cysteine.

If we search for all the proteins of a sequence database containing these patterns, we retrieve all known serine proteases. Therefore, if our putatively orphan protein sequence contains the pattern, it is likely to belong to the chymotrypsin-like serine protease family.

The example that we used is not a hypothetical one; this is indeed the route that was followed to understand the function of the N-terminal region of the NS3 protein of the hepatitis C virus as soon as the sequence of the virus was determined.

5.3 FEATURE EXTRACTION

Some properties of a protein can be computed directly on the basis of its amino acid sequence. We can compute its molecular weight (by simply adding the weights of its amino acids) and its isoelectric point. However, proteins can be subjected to post-translational modifications (glycosylations, phosphoriylations, etc.) and the computed values might not correspond to the experimentally observed values.

Sites where post-translational modifications occur do have specific sequence patterns stored in appropriate databases (for example, PROSITE). Unfortunately, these patterns can be very short and not very precisely defined so that a pattern matching in a protein sequence can give indication about the location of potential modification sites, but often with rather low specificity.

A substantial fraction of proteins are membrane proteins, spanning the membrane or being embedded in it, and they can be identified on the basis of the hydrophobicity pattern of their sequence. A *hydrophobicity plot* of a protein sequence can also be useful to identify the regions buried in the interior of the protein (core). A commonly used method to highlight regions with high hydrophobicity consists in computing the average hydrophobicity value over a "sliding region" or "sliding window" running along the sequence—for example, calculating the average hydrophobicity values of the amino acids in the regions 1–9, 2–10, 3–11, and so on.

Several tables list the hydrophobicity of each amino acid. These "hydrophobicity scales," albeit different, are all sufficiently good to identify clearly hydrophobic and hydrophylic regions and the resulting plots are reasonably well correlated to buried and exposed regions in the three-dimensional structure of proteins, as can be seen in Figure 5.1.

5.4 SECONDARY STRUCTURE: PART ONE

As soon as the first x-ray structures of proteins revealed that they include regions with a repetitive structure, such as α-helices and β-sheets, methods to predict their location started flourishing.

Amino acids occur with different frequencies in α-helices and β-sheets; that is, they show different propensities to be part of the two secondary structure types. This observation forms the basis of many prediction methods, which differ for the

FLAV_CLOBE

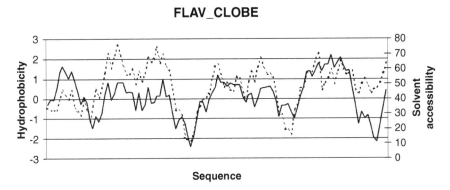

FIGURE 5.1 Comparison between the hydrophobicity plot of the sequence of flavodoxin (continuous line) and the values of the solvent accessible surface area (broken line) computed from its three-dimensional structure.

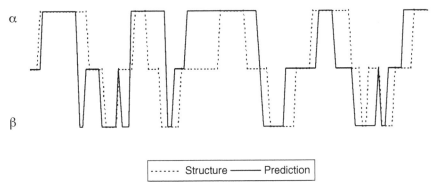

FIGURE 5.2 Comparison between the predicted (continuous line) and experimental (broken line) secondary structure of flavodoxin.

empirical rules that they use (for example, how many amino acids with a high helix propensity must be next to each other in order for a helix to be predicted in the region); for the statistical analysis; and for the use of propensity values derived for a single amino acid or pairs or triplets of amino acids.

The efficiency of all these methods is not very high when a single amino acid sequence is used as input, as can be appreciated by looking at Figure 5.2, where the known secondary structure of flavodoxin is compared with the prediction obtained with one of these statistical methods (the Garnier one, in this case). Other statistical methods based on single sequences also have low reliability, while methods using multiple alignments of proteins as input can reach a higher reliability, as we shall see in the next chapter.

Another method sometimes used to predict the position of α-helices is based on a *helical wheel* representation of the sequence. An α-helix has a pace of 3.6 amino acids. If we assume that one of its sides is exposed to the solvent and the other is

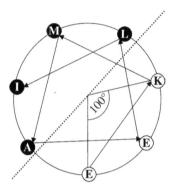

FIGURE 5.3 Representation of a sequence fragment (EKMAELI) on a helical wheel. The amino acids in black circles are hydrophobic; the line divides the putative hydrophobic face from the hydrophilic one.

buried in the protein core (i.e., if we assume that the helix is amphiphilic), we should find hydrophobic amino acids every three or four positions.

A very easy way to verify whether a region of the protein has the periodicity expected for an amphiphilic helix is to draw a helical wheel—that is, project the structure of the putative helix on a plane, with each residue rotated by 100° with respect to the previous one—and verify whether the resulting drawing is compatible with an amphiphilic helix, as shown in Figure 5.3. (One section of the projection plane should almost exclusively contain hydrophobic amino acids and the other hydrophilic ones.)

Orphan sequences are bound to become rarer, thanks to the increase in the number of known sequences and structures. It is likely that their annotation will be simplified when we will be able to inspect the metabolic pathways of completely sequenced organisms and look for missing enzymes. Although it will take some time to complete the assignment of the metabolic pathways for complex organisms, the speed at which new experimental and computational methods keep appearing is very promising.

REFERENCES

Historical Contributions

Chou and Fasman's paper described one of the first methods for secondary structure prediction as well as the derived propensities of amino acids for different secondary structure regions:

Chou, P. Y. and G. D. Fasman. 1978. Prediction of the secondary structure of proteins from their amino acid sequence. *Advances in Enzymology and Related Areas of Molecular Biology* 47:45–148.

Hydrophobicity profiles in proteins can be found in:

Eisenberg, D., R. M. Weiss, and T. C. Terwilliger. 1984. The hydrophobic moment detects periodicity in protein hydrophobicity. *Proceedings of*

the *National Academy of Sciences of the United States of America*
81:140–184.

Suggestions for Further Reading

A description of the methods used to analyze single sequences can be found in:

Banerjee–Basu, S. and A. D. Baxevanis. 2001. Predictive methods using protein sequences. In *Bioinformatics: A practical guide to the analysis of genes and proteins*. Baxevanis, A.D. and Oullette, B.F., eds., New York: Wiley–Liss, Inc.

PROBLEMS

1. The Chou and Fasman method for predicting the presence of a helix is based on the following rules:
   a. Assign to each amino acid a propensity (see following table) for forming a helix.
   b. Search for regions where four out of six contiguous residues have Pa higher than 1.0. The region is temporarily defined as a helix
   c. Extend the helix in both directions until the average Pa value computed for a set of four contiguous residues is lower than 1.0.
   d. If, at the end of these steps, the segment is longer than five residues and the average Pa is higher than the average Pb for the whole region, then the segment is predicted to be helical. Use the method for assigning the predicted location of the helices in the sequence of ubiquitin: MQIFVKTLTGKTITLEVEPSDTIENVKAKIQDKEGIPPDQQRLIF-AGKQLEDGRTLSDYNIQKESTLHLVLRLRGG. Compare the results with the experimentally determined secondary structure of the protein and comment on the result.

Chou and Fasman Preference Parameters for α-Helices (Pa), β-Strands (Pb), and Coils (Pc)

Amino acid	Pa	Pb	Pc	Amino acid	Pa	Pb	Pc
Alanine	1.42	0.83	0.66	Leucine	1.21	1.30	0.59
Arginine	0.98	0.93	0.95	Lysine	1.14	0.74	1.01
Aspartic acid	1.01	0.54	1.46	Methionine	1.45	1.05	0.60
Asparagine	0.67	0.89	1.56	Phenylalanine	1.13	1.38	0.60
Cysteine	0.70	1.19	1.19	Proline	0.57	0.55	1.52
Glutamic acid	1.39	1.17	0.74	Serine	0.77	0.75	1.43
Glutamine	1.11	1.10	0.98	Threonine	0.83	1.19	0.96
Glycine	0.57	0.75	1.56	Tryptophan	1.08	1.37	0.96
Histidine	1.00	0.87	0.95	Tyrosine	0.69	1.47	1.14
Isoleucine	1.08	1.60	0.47	Valine	1.06	1.70	0.50

2. Find the sequence pattern associated with the **LDL-receptor class A (LDLRA) domain signature** and describe it in words.

3. How many times is the pattern found in the *Escherichia coli* proteome?

4. Use the GES hydrophobicity scale (see following table) and plot the hydrophobicity profile of the sequence. How many transmembrane regions would you predict from the plot?

> EAQITGRPEWIWLALGTALMGLGTLYFLVKGMGVSDPDAKK-
> FYAITTLVPAIAFTMYLSMLLGYGLTMVPFGGEQNPIYWARY-
> ADWLFTTPLLLLDLALLVDADQGTILALVGADGIMIGTGLV-
> GALTKVYSYRFVWWAISTAAMLYILYVLFFGFTSKAESMRPEV
> ASTFKVLRNVTVVLWSAYPVVWLIGSEGAGIVPLNIETLLFMV-
> LDVSAKVGFGLILLRSRAIFGEAEAPEPSADGAAATS.

GES Scale for Hydrophobicity of the 20 Amino Acids

Aa	Hydrophobicity	Aa	Hydrophobicity
A	1.6	M	3.4
C	2	N	−4.8
D	−9.2	P	−0.2
E	−8.2	Q	−4.1
F	3.7	R	−12.3
G	1	S	0.6
H	−3	T	1.2
I	3.1	V	2.6
K	−8.8	W	1.9
L	2.8	Y	−0.7

5. Draw the helical wheels for the following two fragments and assess which one is more likely to be an amphiphilic helix:

> HLVGLIDRVFSAAVDFVRVV
> LADFGEDCVHKRFSWSGHIQ

6. Do you think that the enclosed sequence corresponds to a membrane or to a globular protein? Why?

> ALHWRAAGAATVLLVIVLLAGSYLAVLAERGAPGAQLITY-
> PRALWWSVEATTVGYGDLYPVTLWGRCVAVVVMVAGITSF-
> GLVTAALATWFVGREQ

7. Write a flow chart describing the steps you would perform for analyzing a protein sequence.

8. Find how many times the PROSITE pattern PS00124 (fructose-1-6-bis-phosphatase active site) occurs in *E. coli*.

9. Given the table of amino acid usage in *E. coli* and the length of its proteome (about 10^6 amino acids), what is the chance of randomly finding the fructose-1-6-bisphosphatase active site pattern?

Frequency of Amino Acids in the *E. coli* Proteome	
Amino Acid	Frequency (%)
A	9.51
V	7.07
L	10.67
I	6.00
G	7.38
P	4.43
F	3.89
H	2.27
W	1.53
Y	2.84
C	1.16
M	2.82
K	4.41
R	5.52
D	5.15
E	5.76
N	3.95
Q	4.44
S	5.80
T	5.40

10. How much more or less likely would it have been to find the pattern in *E. coli* if the distribution of amino acids in the proteome were uniform?

6 Prediction of the Three-Dimensional Structure of a Protein

GLOSSARY

CASP: critical assessment of methods for protein structure prediction; an international experiment to evaluate the accuracy of methods for structure prediction; methods are tested using a set of proteins before their structure is publicly available

Core: part of the structure that is structurally conserved between two homologous proteins

Random coil: configuration of a protein in the unfolded state

r.m.s.d.: root mean square deviation; defined as: $\sqrt{\dfrac{\sum_i [x_i - x'_i]^2}{N}}$, where x_i and x'_i are the coordinates of the corresponding atoms and N is the total number of superimposed atoms

Test set: set of data not used in the training set, and left aside to assess the reliability of a method

Training set: set of data used to parameterize (or train) a program—for example, an artificial neural network

6.1 BASIC PRINCIPLES

We have mentioned several times that one of the main interests of bioinformatics is to assign a function, possibly at the molecular level, to gene products. In other words, we want to associate each gene to its own product and unravel the molecular mechanism responsible for the function of the latter. Ideally, if we are dealing with enzymes, we would like to identify the groups involved in catalysis as well as those involved in substrate recognition.

Given a genome, there are molecular mechanisms able to recognize where genes are and translate them into functional products. As we mentioned, our understanding of how this happens is incomplete, making it difficult to predict which products are encoded by a genomic sequence, especially in higher organisms.

Once the translation process is complete, a protein assumes a unique three-dimensional structure, by and large determined by its amino acids sequence. The shape of the protein allows it to perform its function. Once again, our knowledge of the process is only approximate and thus we do not know, in general, how to predict the structure and the molecular function of a protein. However, we are able to predict the structure of some proteins (at different levels of accuracy) in some specific cases that are not rare.

Ideally, we would like to be able to reproduce *in silico* what nature does *in vivo*: simulate in a computer the behavior of a polypeptide chain in its physiological environment, following all the steps that lead from a *random coil* (i.e., an unstructured state) to a stable, native, energetically favored, and functional structure. At present, the simulation of the folding process is out of our capabilities, as we shall see in Chapter 9. Nonetheless, we can review what we already know about protein structures and see whether we can extract some rules that can be used to predict structure from sequence, at least in some cases.

A polypeptide chain assumes a single conformation in order to reach a stable, low free energy state. The forces involved are entropic, enthalpic, and kinetic. In other words, a given protein assumes a defined conformation because it minimizes its free energy in the process that leads from a random coil to the native state, via a kinetically allowed path.

A polypeptide chain is made by a polar main chain (the carbonyl [CO] and amino [NH] groups are polar) and by hydrophobic or hydrophilic side chains. In a random coiled state, the polar groups of a protein form hydrogen bonds with the molecules of a polar solvent, while hydrophobic groups, which cannot engage in favorable interactions with the solvent, force the molecules of the solvent to order themselves so as to maximize the number of hydrogen bonds among them. Packing together apolar molecules reduces the number of ordered polar solvent molecules and therefore increases the entropy. This is the reason why oil forms drops in polar solvents.

In summary, a soluble protein folds because it needs to shield its hydrophobic groups from the solvent without losing the energy associated with the hydrogen bonds of its polar groups with the solvent. In other words, a protein must satisfy as much as possible its hydrogen bond potential, while shielding its hydrophobic groups from the polar solvent.

The most efficient way of saturating all hydrogen bond donors and acceptors of the main chain of a protein is to form repetitive structures so that donor atoms are correctly oriented and at the right distance with respect to acceptor atoms. The most frequently observed repetitive structures found in proteins are the α-helices and β-sheets, as we already discussed in Chapter 1. If a protein has to be globular (i.e., compact) so that it can shield its hydrophobic residues within a buried *core*, the main chain needs to turn. This is the structural role of loops and turns.

The question arises then of whether we can predict the location of the secondary structure elements in a protein sequence. This is the subject of this chapter. In the next chapter, we will see how we can exploit another observation that we have already mentioned: proteins with similar sequences evolved from a common ancestor. This implies that mutations, insertions, and deletions accumulated during evolution must have been compatible with the function of the protein. Usually, function is brought about by a few specific residues, but they are held in the correct location by the rest of the structure. Hence, the rest of the sequence is there to guarantee that the protein assumes a stable functional structure. As a consequence, homologous proteins must have similar structures, and this feature is at the basis of a method for predicting the structure of a protein known as homology modeling, which will be described in Chapter 7.

We will also see that the relationship between similarity of sequence and similarity of structure is not bidirectional. While proteins with similar sequences have similar structures (because they are evolutionarily related), proteins with different sequences do not necessarily have different structures. Nature seems to use similar architectures repeatedly in different proteins, even if they perform different functions. Chapter 8 will describe how this observation can be used to build structural models of proteins with a range of methods known as "fold recognition" methods. Chapter 9 will be dedicated to the so-called "new fold" methods aimed at building models of proteins whose structure does not resemble any of the known protein structures.

Before going into the details of the different methods that can be used to infer the structure of a protein from its sequence, we need to address an important problem: How can we evaluate the reliability of prediction methods? The problem of assessing the reliability of a prediction method is far from being trivial and it has been addressed (and at least partially solved) by the scientific community by setting up a worldwide experiment named *CASP* (critical assessment of techniques for protein structure prediction).

6.2 THE CASP EXPERIMENT

The quality of a structural prediction can only be evaluated by comparing the computational model with an experimentally determined structure. This causes a problem because most of the prediction methods are heuristic—that is, based on a collection of statistical observations derived from known structures—and therefore contain information on the experimental structures that can be used to test the method. When the parameters are directly and clearly extracted from known structures, the problem can be solved by parameterizing the methods using all known structures except for a subset that will be used for evaluating its reliability. For example, when we calculate the preference parameters of each amino acid for a given secondary structure element and build a secondary structure prediction method, we use a subset of known structures for deriving the preference parameters, the so-called *training set*. The reliability of the method must be computed on a different set of protein structures, the *test set*, which does not contain the proteins of the training set and, in this specific case, also no protein homologous to any of them, since we know that secondary structure is conserved among homologous proteins.

The use of disjoint training and test set is not always equally straightforward; sometimes the heuristic is more "hidden" and we cannot be sure that the reliability computed on the test set is representative of the reliability of the method when it is applied to a completely new protein. In other cases, the number of available examples can be too small and we need to use the complete set of available data to derive a reliable method. On the other hand, we can only evaluate the performance of a prediction method by comparing its results with experimental facts.

One possible solution to the problem is to predict the structure of an unknown protein the structure of which will be solved shortly after the prediction has been made. This is the spirit of the CASP experiments, which work as follows:

- Every 2 years, crystallographers and NMR spectroscopists who are about to solve a protein structure are asked to release the sequence of their target protein to the scientific community.
- The sequences are made available to the world, together with the date when the structure is expected to be experimentally solved.
- Predictors deposit their predictions before the experimental structure is solved.
- As soon as the experimental structures have been solved and made available, a panel of experts (the assessors) compares models and structures, derives general conclusions on the state of the art, and assesses the quality of the results obtained by the different tested methods. There are three assessors. One evaluates predictions of proteins homologous to proteins of known structures, one those deposited for proteins sharing a structural similarity but no clear sequence relationship with a protein of known structure, and one all the remaining cases. There have been recent changes in the details of the structure of the experiment, but they do not affect what we will discuss next.
- Finally, the results are presented and discussed at a conference where assessors and predictors convene and subsequently published on the Web and in a special issue of the scientific journal *Proteins: Structure, Function and Bioinformatics*.

There have been CASP experiments in 1994 (CASP1), 1996 (CASP2), 1998 (CASP3), 2000 (CASP4), 2002 (CASP5), 2004 (CASP6), and so on.

Throughout this and the next three chapters, we will use the CASP results to give a feeling of the level of accuracy of the various structure prediction methods. It is therefore important to discuss some of the aspects and limitations of this experiment, which has somehow changed the field of structure prediction evaluation.

There are three players in CASP experiments: predictors, assessors, and the final users of the methods. Predictors are not in the ideal situation to produce the best possible models since, for its very nature, CASP poses a time limit. When the experimental structure of a target becomes available, all predictions must have already been deposited. In practice, this implies that predictors need to submit models for all or most target proteins (which can be close to a hundred) in a few weeks. This does not reflect what predictors do in real life, where they can continue to refine and analyze the model for as much time as needed.

Furthermore, the results and performance of each method are made publicly available and this might discourage the application of new "bold" methodologies. Moreover, the number of targets in each category is limited, especially as far as new folds are concerned, and the results might, in some cases, lack solid statistical bases. Finally, there is a subjective element in the evaluation due to the choices and opinions of the assessors.

This latter observation might sound odd. Given an experimental structure and a model of the same protein, how can the result of a comparison be subjective? In practice, the evaluation cannot be but subjective because it depends upon what we consider a good quality model. For example, is it better to predict the complete structure of a target or is it more sensible only to produce a model for those regions that can be modeled with sufficient accuracy? Is it better to produce a model with an overall good quality or a model in which part of the structure is predicted with extremely good quality at the expense of an average quality of the rest of the structure?

There are numerical values that can be computed to assess the quality of a model; for example, we can compute the *r.m.s.d.* between the experimental structure and each model after optimal superposition (but be careful because the solution is not unique). But which atoms will we superimpose—all the atoms, only the main chain atoms, only the atoms of the main chain and of the buried side chains, only those in the core, those of the biologically important regions, etc.? How do we decide whether a specific target was more or less difficult to model than others? How do we take into account experimental errors and uncertainties?

Each assessor has his or her view about what is really relevant to the final users and this does introduce a subjective element. This notwithstanding, the CASP experiments have revealed to be extremely useful in pushing the prediction field and in identifying the bottlenecks of prediction methods.

6.3 SECONDARY STRUCTURE PREDICTION: PART TWO

We have already briefly described methods for secondary structure prediction based on a statistical analysis of the preference of each amino acid for a given secondary structure element. We have mentioned that their reliability is quite limited when they only use the information on a single amino acid sequence.

On the other hand, we do know that members of a protein family have a similar three-dimensional structure and therefore also a similar secondary structure pattern. Therefore, we can use the information derived from a multiple-sequence alignment of a family to predict the secondary structure of its members. In other words, the input of a prediction method can be a multiple-sequence alignment or a profile rather than a single sequence, provided we can find a suitable way of encoding it.

As we mentioned, a multiple-sequence alignment of proteins that are N amino acids long can be represented by a matrix (i,j), called a profile, where the index i represents the position along the sequence (hence, it goes from 1 to N) and the index j represents 1 of the 20 amino acids (hence, it goes from 1 to 20). The cell i,j will contain the number of times the amino acid j is present in position i of the alignment (Figure 6.1).

This matrix can be used as input for a neural network. Usually, the input of the network is the profile of a running window that slides along the alignment and the prediction will refer to the central amino acid. In this way, we can take care of "context effects" (i.e., of the contribution of the local sequence in determining the

	A	C	D	E	F	G	H	I	K	L	M	N	P	Q	R	S	T	V	Y	W
GGGGG						5														
YYYYY																			5	
IIILL								3		2										
VVVVL										1										
SSTTT																	4			
TSTSS																3	2			
RRRRR															5					
KKKDD			2						3											
ILIFL					1			2		2										
SSTGH						1	1									2	1			
VIVVV								1										4		
GGGGG						5														
KKKKK									5											
PPPPP													5							
LLLIL								1		4										
FYWWW					1														1	3
LILIL								2		3										
TRSRT															2	1	2			
AAALI	3							1		1										
VVVVV																		5		
KKKKK									5											
RRRRR															5					

FIGURE 6.1 A possible way of encoding a multiple alignment.

secondary structure). If we use a window of length 13, the prediction will refer to the central (seventh) amino acid, but it will also use information about the alignment of the preceding and following six positions (Figure 6.2).

The network will need to be trained and, to this aim, we can subdivide the database of proteins of known structure into two sets. One will be used to train the network and the other to test its performance. We should make sure that no protein of the training set is homologous to any protein of the test set, since in this case we might end up training the network in recognizing homology rather than in predicting secondary structure.

Methods for protein secondary structure prediction based on multiple sequence alignment and automatic learning methods, such as neural networks, can reach an accuracy of up to 80%. These values are always averages over a number of test cases; the accuracy of the prediction will be higher for proteins for which a good multiple sequence alignment is available and lower if this is not the case. Since these

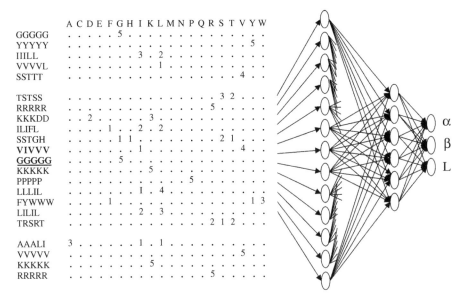

FIGURE 6.2 A schematic representation of an artificial neural network for the prediction of secondary structure. The prediction for the central residue of the window (underlined) will depend upon which output cell has the highest value. In this example, the prediction could be α-helix (α), β-sheet (β), or loop (L). The window will slide by one position at the time along the sequence.

methods have reached such a respectable accuracy, we can start addressing the next logical step: how to use the approximate knowledge of the location of secondary structure elements in a protein together with some information about long-range contacts between amino acids to try to assemble a full-fledged three-dimensional model of the protein.

6.4 LONG-RANGE CONTACT PREDICTION

At least in principle, a multiple-sequence alignment of a family of proteins can also provide some information about long-range contacts. All members of the protein family have the same fold; therefore, pairs of amino acids in contact in one protein will be aligned to pairs of amino acids in contact in the other members of the family. If, during evolution, one of the amino acids of the pair has undergone a mutation, it is possible that the contacting one will have experienced a compensating mutation. For example, if the amino acid in a position of one protein has been replaced by a larger one in another protein, the contacting one might have been replaced by a smaller amino acid. Of course, we are making the simplifying assumptions that each amino acid is only in contact with another one and that the local main chain structure does not change at all.

In these hypotheses, by analyzing a multiple-sequence alignment, we might be able to identify pairs of positions that change in a correlated fashion during evolution. (If the amino acid in one position changes, in a different protein, from a large to a small amino acid, the other will change from a small to a large one and vice versa, or the change in one position of a positively charged amino acid with a negatively charged one might correspond to a change from a negatively charged to a positively charged one in a contacting position.)

There are methods that attempt to highlight this correlated behavior of amino acids in order to identify positions putatively in contact in the three-dimensional structure, but as we mentioned, they rely on rather strong assumptions, which limits their reliability. Amino acids are not just in contact with one other amino acid, and compensative changes can be redistributed among several residues. Furthermore, proteins are not rigid objects and small movements of the main chain or of neighboring side chain can allow a larger amino acid to be accommodated in the structure without necessarily involving correlated mutations.

Nevertheless, in some cases, strongly correlated positions can be identified and used, for example, to decide between alternative possible topologies for the protein. Possibly in the future, these methods might even allow exploitation of secondary structure predictions for folding up proteins.

6.5 PREDICTING MOLECULAR COMPLEXES: DOCKING METHODS

Proteins rarely act alone; often their action is brought about by complexes. The problem arises, therefore, of being able, given the structure of two interacting proteins, to predict the structure of their complex.

This problem is not trivial and can be addressed using energetic methods (two proteins form a complex because its formation is energetically favored) with geometrical methods (the interacting surfaces need to be complementary) or a combination of the two strategies.

The general considerations about the accuracy of energetic evaluations that we will discuss in Chapter 9 and the large number of possible orientations that need to be explored represent major obstacles towards the solution of the problem. We also have to face the problem that the structure of a protein in a complex is not exactly the same as that of the protein by itself. It could not be otherwise, since the protein is in a different environment in the two cases and experiences different interactions.

It is also difficult to evaluate the effectiveness of the available methods correctly. An experiment, named CAPRI and very similar to CASP, is ongoing. The number of cases that can be used for "blind" tests—that is, before the structure of the complex is experimentally solved—is still rather limited, so it is difficult to derive general conclusions.

Docking programs almost never give one single solution; more often they provide the user with a list of ranked solutions, often including, in some position, a solution very close to the correct one. These lists are very useful in any case, since they might allow design of appropriate experiments to find the correct mode of interaction.

A different and more general problem, albeit much more difficult to solve, is the prediction of whether and how two proteins interact on the basis of their sequences. With the data deluge provided by genomic projects, a method for solving this problem would be really invaluable. We are far from solving the problem in an effective and general way, but there have been very interesting attempts. Some methods use the same approach that we described to find long-range contacts (i.e., correlated mutations). In this case, the correlation is analyzed between positions of multiple alignments of two different protein families rather than within the sequence of a single protein family.

Another strategy consists of exploiting the observation that two proteins that interact in one organism might be homologous to two domains of a protein of another organism. When this is the case, it is indeed very likely that the two proteins interact.

Now that the number of sequenced genomes is respectably high, some methods use patterns of presence or absence of genes for identifying interactions. If two genes are always present or always absent in a set of genomes, it is likely that their products are part of the same biological process and they are therefore candidates for being physically interacting.

A related, albeit different, problem to that of protein docking is the prediction of the interaction of a protein with small molecules (substrate, inhibitor, cofactor). This is of clear applied interest and indeed there is a plethora of available methods, which we will not discuss here. However, it is worth mentioning that, in this case even more than in the case of protein–protein docking, the problem of predicting the conformational changes occurring upon binding is important. The structure of a small molecule is even more sensitive to differences in its environment and interactions. If the small molecule does not have too many degrees of freedom, we can try docking it using several conformations. More often the problem is addressed in a "differential" form, when we need to predict the interaction of a protein with a series of reasonably similar small molecules. If we know the mode of binding of one of the molecules of the series, we can assume that the others will bind in a similar fashion.

These methods are routinely applied in every research center of pharmaceutical companies and, even if it is still the exception rather than the rule that a drug is "designed" at the computer, docking calculations have certainly accelerated substantially the drug discovery process and have become standard tools in pharmaceutical chemistry.

REFERENCES

Historical Contributions

A milestone in protein structure prediction history can be found in:

Levinthal, C. 1966. Molecular model-building by computer. 1966. *Scientific American* 214:42–52.

A revolution in the prediction of secondary structure elements came from the combination of multiple alignments and artificial neural networks:

Rost, B., and C. Sander. 1993. Prediction of protein secondary structure at better than 70% accuracy. *Journal of Molecular Biology* 232:584–599.

The first method for predicting long-range contacts, based on multiple alignments, was developed by:

Gobel, U., C. Sander, R. Schneider, and A. Valencia. 1994. Correlated mutations and residue contacts in proteins. *Proteins: Structure, Function and Genetics* 18:309–317.

Suggestions for Further Reading

The CASP experiments are described in a special issue of the journal *Proteins, Structure Function and Genetics*. The most useful papers are the ones in the introduction and those of the evaluators (one per category). A few examples are listed here, but of course the list grows every 2 years:

Moult, J., J. Pedersen, R. Judson, and K. Fidelis. 1995. A large-scale experiment to assess protein structure prediction methods. *Proteins: Structure, Function and Genetics* 23:ii–v.

Moult, J., T. Hubbard, et al. 1997. Critical assessment of methods of protein structure prediction (CASP): Round II. *Proteins: Structure, Function and Genetics* Suppl 1:2–6.

Venclovas, C., A. Zemla, K. Fidelis, and J. Moult. 1999. Some measures of comparative performance in the three CASPs. *Proteins: Structure, Function and Genetics* Suppl 3:231–237.

Docking methods have been reviewed in only one of the CASP experiments (CASP2):

Dixon, J. S. 1997. Evaluation of the CASP2 docking section. *Proteins: Structure, Function and Genetics* Suppl 1:198–204.

Hart, T. N., S. R. Ness, and R. J. Read. 1997. Critical evaluation of the research docking program for the CASP2 challenge. *Proteins: Structure, Function and Genetics* Suppl 1:205–209.

Kramer, B., M. Rarey, and T. Lengauer. 1997. CASP2 experiences with docking flexible ligands using FlexX. *Proteins: Structure, Function and Genetics* Suppl 1:221–225.

Sobolev, V., T. Moallem, et al. 1997. CASP2 molecular docking predictions with the LIGIN software. *Proteins: Structure, Function and Genetics* Suppl 1:210–214.

Some papers about docking methods based on geometry and/or molecular dynamics include:

Di Nola, A., D. Roccatano, and H. J. Berendsen. 1994. Molecular dynamics simulation of the docking of substrates to proteins. *Proteins: Structure, Function and Genetics* 19:174–182.

Hart, T. N., and R. J. Read. 1992. A multiple-start Monte Carlo docking method. *Proteins: Structure, Function and Genetics* 13:206–222.

Helmer–Citterich, M., and A. Tramontano. 1994. PUZZLE: A new method for automated protein docking based on surface shape complementarity. *Journal of Molecular Biology* 235:1021–1031.

Vakser, I. 1997. Evaluation of GRAMM low-resolution docking method-ology on the hemaglutinin–antibody complex. *Proteins: Structure, Function and Genetics* Suppl 1:226–230.

PROBLEMS

1. Given a set of proteins of known structure, describe which filtering pro-cedure you could use to select the training and test set for an automatic learning method for predicting the secondary structure of a protein.
2. A measure of the accuracy of secondary structure is called Q3 and is defined as the percentage of correctly predicted alpha, beta, and "other." Compute its value for the following example:

Experiment	H	H	H	H	C	C	C	E	E	E	E	C	C	E	E	E	E	C	C	H	H	H	H	C
Prediction	C	H	H	H	H	C	C	C	E	E	E	C	C	E	E	C	E	E	C	H	H	C	H	H

3. Several strategies can be used to extract secondary structure elements automatically from a PDB file. Describe a possible method.
4. Explain in a few words how a layered neural network is trained.
5. Count the number of adjustable parameters for the neural network shown in Figure 6.1P.
6. Explain why sparse encoding is good a way to process (encode) a protein sequence into a neural network. Discuss the encoding used in the PHD program for predicting secondary structure.
7. Given the following multiple sequence alignment, could you identify some positions that might be evolving in a correlated way? Ignore the fact that the number of sequences is too low to derive statistically sound conclusions!

A	D	D	E	F	C	G	S	T	E	R	D	S	T	G	W	A	C	I	T
A	D	D	D	F	C	G	S	T	E	R	D	S	T	G	W	A	C	I	T
F	D	R	D	F	L	G	S	R	E	E	D	S	S	G	V	A	L	L	T
F	D	R	T	F	L	G	S	A	E	E	D	S	S	G	V	A	L	V	T
A	D	R	E	F	I	G	S	T	E	E	D	S	T	G	W	A	L	L	T

8. Which conclusions could you derive from the alignment of problem 7? Which information about the protein would you need to draw more con-clusions?
9. In the CAPRI experiment, one way to evaluate the predictions is to compute how many contacts between two partners of the complex structure to be predicted are correctly identified. Describe how you would derive sensitivity and specificity values for a prediction.

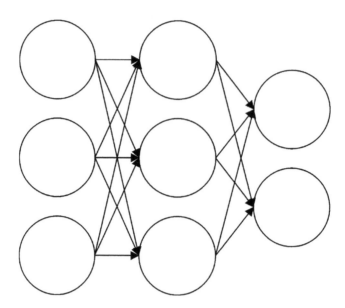

FIGURE 6.1P Scheme of a two-layered neural network.

10. Which other parameters do you think could be used to evaluate the correctness of a docking result?

7 Homology Modeling

GLOSSARY

Alignment: correspondence between the amino acids of two or more proteins

Homology modeling: technique for the prediction of the three-dimensional structure of a protein that relies on the possibility of using the structure of a homologous protein as starting model

Loop: region with nonrepetitive structure (i.e., not alpha or beta)

Rotamer library: list of the most frequently observed dihedral angles for each amino acid side chain

Sequence identity: percentage of identical amino acids between two aligned sequences

Sequence similarity: score of an alignment obtained by adding the similarity values for each pair of aligned amino acids. The similarity value of a pair of amino acids can be derived taking into account their chemical and physical similarity, the frequency with which they replace each other during evolution, or the minimal number of DNA base replacements needed to change the codon codifying for one in a codon codifying for the other.

Solvent accessibility: surface of an amino acid that is accessible to molecules of the solvent

Stems: regions flanking an insertion or a deletion of one or more amino acids in a protein structure

Target protein: the protein to be modeled

Template protein: the protein homologous to the target protein, the structure of which will be used as starting approximation for building the model

7.1 BASIC PRINCIPLES

Often a biological project can benefit from the knowledge, even approximate, of the three-dimensional structure of a protein. If the structure of the protein has not been experimentally determined, one can try to build a model of the protein of interest (*target protein*). In order to do so, the first step to be performed is a search in the database of solved protein structures to verify whether a protein of known structure (*template protein*) with a similar amino acid sequence exists. By now, the reasons for doing so should be clear, but let us recall them once more.

We expect that two homologous proteins (i.e., two proteins deriving from a common ancestor) have a similar structure since only nondestabilizing mutations are accepted by evolution. On the other hand, the amino acid sequences of two homologous proteins are expected to share a similarity because they come from the

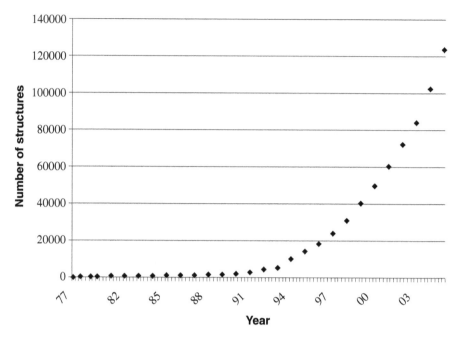

FIGURE 7.1 Growth of the structures deposited in the PDB database.

same original sequence (and the closer the evolutionary relationship is the more similar will be the two sequences). It follows that we can hope to detect homology, and therefore structural similarity, between two proteins on the basis of the similarity of their sequences

In practice, there is a relationship between the similarity of two protein sequences: the number of similar or identical amino acids between them is related to the similarity of their structures. As we have mentioned and as we will discuss in the next chapter, the lack of a significant sequence similarity does not imply that the structures are different. The similarity between the structure of the main chain of two homologous proteins increases, in a predictable way, as the similarity between their sequences increases. In practice as we will see, this is the case for the "core" regions of the proteins and does not apply to exposed, flexible regions (i.e., to the loops).

On the basis of this relationship, we can build a three-dimensional model of a protein taking advantage of a technique called *homology modeling*, which is without doubt the most accurate and effective way to predict the structure of a protein when the structure of one of its homologous proteins is known. The homology modeling method is based on heuristic observations; that is, it takes advantage of available structural data to derive the rules that will be employed for building the model, rather than attempting to simulate the natural folding process of a protein.

The speed at which new structures are determined by nuclear magnetic resonance (NMR) and x-ray crystallography is impressive (Figure 7.1). Nevertheless, for very good reasons, the interest for prediction methods, especially homology modeling, is not fading. It is estimated that the proteins that can be built by homology as of today are at least 10 times more than those of known, experimentally determined structure.

The aim of this chapter is to describe this modeling technique, highlighting its advantages and pitfalls so that the reader can critically evaluate its results. This is of paramount importance since, as we shall see, each of the steps in the procedure introduces errors that are not uniformly distributed along the models.

In homology modeling it is possible to evaluate the expected reliability of a model beforehand, since it essentially depends upon the degree of similarity between the target and template proteins. It is therefore possible to decide a priori whether the model is sufficiently reliable for the specific desired application. In most cases, a homology model can be used to predict which mutations are likely to be compatible with the overall structure of the protein. More in-depth information, such as the values of the *solvent accessibility* of a side chain or details about its packing within the protein, can be derived from a model only if it is based on a template protein of known structure with sufficiently high *sequence similarity* (i.e., sufficiently close in evolutionary terms). Finally, it is unreasonable to use almost any model for detailed energetic calculations, or to quantitatively predict the affinity of the protein for a ligand, or to design such a ligand.

With the only possible exception of models based on a very high *sequence identity* with their structural template, the only one way to increase the confidence in the reliability of a model is to use it to predict properties of the target protein that can be experimentally verified.

7.2 THE STEPS OF COMPARATIVE MODELING

Homology modeling is based on the empirical observation that homology between two protein sequences that can be detected on the basis of their sequence similarity implies structural similarity. It follows that the coordinates of the main chain of the template protein can be used as a first approximation of the coordinates of the corresponding regions of the target protein (according to their sequence *alignment*). A classical homology modeling approach consists of several steps:

- identification of the proteins of known structure that will be used as a template
- identification of the regions expected to be structurally conserved between the target and the template
- construction of a sequence alignment between the sequences of these regions
- construction of a first approximate model for the main chain of the conserved regions, using as coordinates those of the template protein atoms according to the alignment
- construction of a model of the structurally variable regions (e.g., regions where insertions and deletions are present)
- modeling of the side chains of the target protein
- refinement of the model

7.2.1 TEMPLATE SELECTION

The similarity of the main chain conformation in the core of two homologous proteins increases as their sequence similarity increases. In detail:

- For proteins with a sequence identity above 50%, the r.m.s.d. of the atoms of the main chain is lower than 1.0 Å and the core includes about 90% of the structure.
- For proteins with sequence identity lower than 20%, the core region might represent less than 50% of the structure with an r.m.s.d. of the main chain greater than 1.8 Å. Outside the core, important structural variations can be observed.
- Pairs of proteins with sequence identity between 20 and 50% will have an intermediate degree of structural similarity.

From these admittedly simplified rules of thumb, it follows that the best template to build a model is the protein of known structure with the highest sequence similarity with the target protein. When more than one protein of known structure with a similar sequence identity is available, it is advisable to use other criteria for the selection—for example, the quality of the structural data (the resolution) or its completeness. A typical example of such a case is that of immunoglobulins. There are so many known structures of immunoglobulins and the degree of similarity between their sequences is so high that these factors could and should be taken into account.

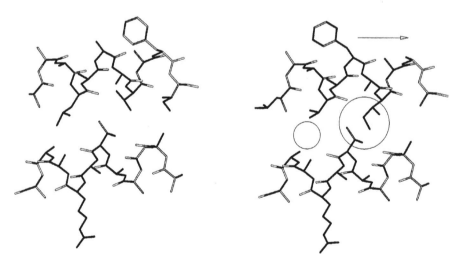

FIGURE 7.2 Example of a case where an alignment error is highlighted by problems in the corresponding model. The larger circle indicates two residues whose side chains are too close to each other, while the smaller one shows a region where the packing of the side chains is too loose. Shifting the sequence by three residues (as shown on the left) can solve both problems.

However, in general, the degree of sequence similarity is not constant along the sequence and, if there is more than one possible template, one could select different templates for different regions (e.g., on the basis of the local sequence similarity). It is unclear whether this is always the best choice. In CASP, methods using multiple templates seem to be, on average, better than those using single templates, although there are significant exceptions.

7.2.2 SEQUENCE ALIGNMENT

So far in this chapter, we have assumed that the sequence similarity between two sequences is easy to measure, taking advantage of the alignment algorithms previously described. Let us recall that the final aim of alignment algorithms is to find the alignment that maximizes the identity (i.e., the number of identical aligned amino acids) or the similarity (i.e., a score reflecting the degree of similarity between aligned pairs of amino acids) between two protein sequences. In the case of homology modeling, we could use them to obtain the likely amino acid correspondence between the target and template sequences.

The algorithms described in Chapter 4 with some of the described approximations, can be used for searching the protein sequence databases for proteins with a sequence similar to that of our target protein and for giving us a measure of the identity of or similarity between their sequences. The latter is usually sufficiently accurate to allow us to judge whether one protein is a suitable template for building a homology model of our target molecule. However, are we certain that the alignment that maximizes the identity or similarity between two sequences always corresponds to the alignment that gives us the best structural superposition? Obviously, the latter is what we need to build a correct first approximation model of our protein.

There is no risk of overstating the importance of the alignment in the model-building procedure; even a small mistake at this stage can produce devastating effects on the final model. On the other hand, there is a conceptual difference between the optimal sequence alignment and a structurally correct alignment. The former is aimed at reproducing the evolutionary events of the protein family and the latter implies that corresponding amino acids have retained their relative position in the protein structures.

For a simple example, the following alignment, although corresponding to the one maximizing the sequence identity between the two polypeptides, cannot correspond to a correct structural alignment:

```
sequence 1  LADGTRCTGRGSDW

sequence 2  LVD-SKCRAKG-DW

            *   *       *   *
```

The amino acids indicated by the stars cannot be structurally equivalent because it is stereochemically impossible that two amino acids are in the same relative spatial position in two structures while they are separated by one amino acid in one of them and none in the other. This is a trivial example, but it is useful to stress the concept that the optimal sequence alignment is a one-dimensional result that needs to be translated into a three-dimensional concept.

A more correct, but rarely used, way to write the preceding alignment would be, for example:

```
sequence 1 LADGT--RCTGR--GSDW

sequence 2 LV---DSKCRAKGD---W
```

This indicates that there cannot be a correspondence between the amino acids on either side of an insertion, it is very important to try to imagine the structural implications of a protein sequence alignment. CASP2 gives us a more realistic example of what we are discussing (Samudrala and Moult 1997).

The target protein for the experiment is the endoglucanase I (the target code in CASP2 was T0028), which shares a 47% sequence identity with a protein of known structure (whose protein structure database code is 1celA). Automatic methods produced the following alignment in the region between amino acids 49 and 70 of the target protein:

```
TARGET:  CTVNGGVNTTLCPDEATCGKNC

           |          |||||   |   |||

PARENT:  CYDGNTWSSTLCPDNETCAKNC
```

Once the structure of the target protein was experimentally solved, it was possible to superimpose target and template structurally. The corresponding alignment was different from the sequence-based one:

```
TARGET:  CTVNGGV----NTTLCPDEATCGKNC

           |                |       ||

PARENT:  CYDGNTWSSTLCP---DNETCAK-NC
```

In other words, the underlined regions of the two proteins cannot be structurally aligned (their main chain differs by more than 4 Å) even though the sequence alignment produces a convincing result in terms of maximizing the number of corresponding identical amino acids.

The alignment that will be used to build the model is obtained by using the amino acid sequences of the two proteins, but we should not forget that we have other data available. Certainly we know the structure of the template protein, most likely the sequences of other proteins of the same family. If more than one protein of the family is of known structure, we can perform a structural superposition between members of the family. It is also rather common that we know of experimental results on one or both proteins. None of this information should be neglected; let us see how we can exploit it.

Insertions and deletions, which are the most difficult regions to model, occur much more frequently on the surface of a protein, where they only cause local structural rearrangements, than in its well packed core or within secondary structure elements. In the latter case, they are much more likely to perturb the structure and/or the function of the protein. Therefore, the template structure can be used to make sure that the alignment algorithm did not locate insertions and deletions in secondary structure elements or in buried regions of the protein. It is also important to verify that the alignment does not imply the presence of non-neutralized charges within the protein core.

The structural superposition of proteins belonging to the family of the target and template protein can instead tell us which regions tend to be structurally conserved in the family and that we can assume are conserved also in our target protein. Equally useful is the multiple-sequence alignment of proteins of the family, since it contains information about the degree of conservation and variability at each of the aligned positions and can therefore help very much in defining the precise location of insertions and deletions.

Any experimental information about the target and template proteins or about any of the members of their evolutionary family should be analyzed and used, whenever possible. If the target and template proteins have a similar or identical function, active site amino acids should be aligned and conserved, epitopes of conformational antibodies should be exposed, and the alignment should be consistent with the results of site-specific mutagenesis experiments.

In an area as complex as protein structure prediction, we cannot afford to discard any information, whatever its type and source. Even after all preliminary sequence alignment steps have been performed and the initial model is being built, it is still worth keeping a critical eye on the sequence alignment. For example, some regions might seem impossible or very difficult to model because there is not enough room to position all the side chains, because the model ends up containing too many cavities, or because the number of amino acids in a loop is inconsistent with the position of its predicted end points. In these and other similar cases, the best solution is to go back to the alignment and try to see whether its modification, possibly manual, can help solve the problems (Figure 7.2).

7.2.3 LOOPS

By definition, regions that are structurally different between the target protein and its structural template cannot be built by homology, so we need to recur to different methods. In general, these parts correspond to *loops*—that is, to regions connecting elements of repetitive secondary structure, such as α-helices or β-strands. They are usually exposed on the surface of the protein, are much less regular than helices and strands, and often contain insertions and deletions. The prediction of their structure is therefore a very difficult problem, but nevertheless a very important one because they often have a functional role. Thanks to their being located on the surface of the protein, they are often involved in protein–protein interaction (e.g., antibodies recognize their cognate antigen in a site formed by six loops very variable in their amino acid sequence) and, in some cases, can be the nucleation site for the folding of the polypeptide.

The methods used to predict the structure of loops can be based on one of three sources of information (or their combination): the sequence of the loop, database search techniques, and energetic calculations. Methods based on the loop amino acidic sequence work fairly well for short loops (three to four residues), especially when the loops connect contiguous β-strands of an antiparallel β-sheet. In this case, as we discussed in Chapter 1, the dihedral angles of the loops need to be such that the chain can invert its direction by 180° within the distance of the hydrogen bonds that link the two connected strands. This implies that some dihedral angles have to assume "unusual" values—values energetically less than optimal and therefore usually assumed by specific amino acids such as glycine. The geometry of the loop dictates in which of its positions such dihedral angles are required and therefore some sequence pattern can be identified, For example, four amino acid long β-hairpins of type I or II usually present a glycine in their third and fourth positions, respectively. Similar preference rules are observed for other types of loops. Unfortunately, these rules are not as useful as they can seem at first sight, since they could be used only if one could know a priori the type of the loop to be modeled.

By definition, all sequence-based rules for predicting the structure of loops are based on the assumption that the conformation of the loop is determined by local interaction and that long-range interactions do not play any relevant role in selecting the loop structure, This is not necessarily true and in some known cases it is clear that the effects of tertiary interactions predominate over the short-range effects (Tramontano et al. 1990).

The problem of predicting short loops is therefore difficult. Even more difficult is the prediction of the conformation of medium sized loops; the number of possible combinations of the dihedral angles of these loops is much higher and their rigorous classification is practically impossible. The only route to their classification is to use the interactions that stabilize them: hydrogen bonds for loops forming compact substructures and hydrophobic packing for loops with a more extended conformation. However, the required interactions can be provided by the rest of the protein structure in many different ways. It follows that this property of loops cannot be used for deriving rules to predict their structure.

One of the most commonly used techniques to model loop regions containing insertions and deletions with respect to the structural template used for modeling is to search the database containing known protein structures for loops that are likely to be good local templates. The rationale for this technique is based on the observation that regions of similar conformation can be found in pairs of homologous and nonhomologous protein structures, as if proteins could repeatedly use fragments of similar conformation, in a sort of protein Lego structure where similar parts can be used to obtain different final objects. But, how can we identify which specific piece is part of our target protein?

One line of reasoning is that, given the predicted structure of the *stems* (i.e., of regions flanking the loop that we need to model), the number of ways to connect them with a given number of amino acids cannot be infinite. Therefore, we can search in the protein structure database for examples of loops connecting stems similar to those observed in our target structure and containing the appropriate number of amino acids. In practice, the method consists of searching the database of solved protein structures for regions similar to the stems that we have modeled and containing the number of amino acids that we need to insert between them (Figure 7.3).

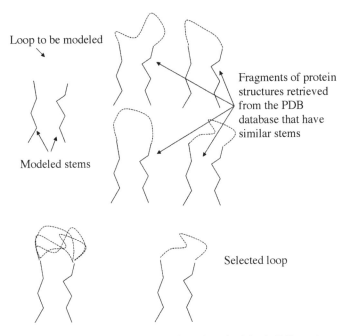

Loop to be modeled

Fragments of protein
structures retrieved
from the PDB
database that have
similar stems

Modeled stems

Selected loop

FIGURE 7.3 Exemplification of the fragment-based method for building a loop.

For the method to work, we need to demonstrate three things. First, that a loop with a conformation similar to the one for which we are searching does exist in the database; second, that similar loops connect similar stems; and, finally, that in the latter cases the geometric relationship between stems and loops is conserved.

As we said, similar loops do exist in different structures, so the first assumption is certainly valid. Unfortunately, this is not the case for the other two hypotheses. In reality, while similar fragments are found in homologous and nonhomologous proteins, the structural relationship between the loops and the stems that they connect is only conserved (but not always) among homologous proteins.

Energy-based methods for predicting the structure of loops are based on com-binatorial searches for possible connecting regions and their subsequent evaluation on the basis of energetic calculations. The energetic evaluation clearly needs to be performed for the loop in the context of the complete protein structure, since we know that identical peptide sequences can appear in different protein structures with very different conformations—another way of saying that nonlocal interactions are responsible for the loop conformations. We will discuss energetic calculations and energy optimization techniques in more detail in Chapter 9, but it is important to here mention here that our ability to compute the energy of interatomic interactions is neither sufficiently complete nor exact to allow us to compute the minimum energy conformation of a protein or of a fragment of a protein. It follows that these methods can work sometimes, but do not work all the time, and it is impossible to establish a priori which is the case for the protein-modeling experiment at hand.

Research in this area is very active, especially because the prediction of the structure of the loops is one of the major stumbling blocks of homology modeling techniques and the place where errors are more frequent and serious. Some methods used to predict the structure of proteins that do not share any sequence or structural similarity with proteins of known structures (described in Chapter 9) are now being applied to the prediction of loop structures.

From a practical point of view, however, it is advisable to critically evaluate whether it is really important to model each of the loops of our target protein. Although some of them might be essential for the protein function, others might be far from the regions we are interested in (for example, the binding or catalytic site); therefore, in many cases an incomplete model lacking some loop regions can be sufficient for our purposes.

7.2.4 SIDE CHAIN MODELING

Once we have built the main chain of the core of our target protein using the coordinates of the corresponding template atoms and we have somehow modeled the loops, we are left with the task of positioning the side chains of our protein. This is a combinatorial problem, and we should, in theory, analyze every possible conformation of each amino acid side chain to find the best combination according to some energetic or packing criterion.

However, the side chains of each amino acid have preferential conformations (Figure 7.4), as can be appreciated by surveying the database of known protein structures and tabulating the frequency at which each side chain is observed in a given conformation. The corresponding dihedral angles are stored in the so-called *rotamer libraries*, which contain, for each amino acid, a sorted list of the most frequently observed dihedral angle combinations; in some cases, these are separated according to the secondary structure to which the amino acid belongs. By using these values, the complexity of the combinatorial problem can be greatly reduced.

Since template and target protein are evolutionarily related, we can also expect that the side chains of the two proteins can assume similar conformations and therefore that the conformations of the side chains of the template are a good first approximation for the position of the side chains of the target, Some methods, indeed, "copy" the dihedral angles of the side chain of the amino acids of the template, as far as the respective length permits, and use rotamer libraries for the remaining part.

Both these strategies are reasonably effective when used for reconstructing the side chains of an experimental structure. Their accuracy decreases as the error on the modeling of the main chain increases. This implies that any improvement in modeling the main chain will provide us with the bonus of better quality of the side chain prediction.

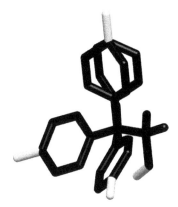

FIGURE 7.4 The most common conformers of the amino acid tyrosine.

7.2.5 MODEL OPTIMIZATION

The steps that we described so far usually allow a reasonably complete model of the protein under study to be built. The next step consists in visually analyzing the model, as well as using specific tools, and evaluating and optimizing it.

In principle, one should fix stereochemical errors, if any, especially in those regions where we have inserted or deleted amino acids, unfavorable interactions, or parts of the model where the packing is clearly suboptimal. This can be done with some manual intervention using a molecular graphics package, followed by a limited number of cycles of energy minimization (see Chapter 9) or of geometric optimization. It is important to stress that neither energy minimization nor geometry optimization can substantially modify a model and, therefore, cannot correct major errors. It has been repeatedly demonstrated that, so far, none of these techniques can consistently improve models; in fact, most of the time they make it worse.

Several cycles of energy minimization, with whatever methods, not only do not improve the model, but also have another disadvantage of "hiding" regions of the model that have problems and are therefore likely to be incorrect. When a model is subjected to intensive energy minimization or optimization, the programs tend to modify the conformation of regions near those where the interactions are unfavorable, in order to lower the overall energy of the model. Thus, at the end, all parts of the model will look equally "nice" and it will become impossible to single out regions that are likely to be incorrect.

7.3 ACCURACY OF HOMOLOGY MODELS

How much can we trust a model? As we said, the quality of a model essentially depends upon the closeness of the evolutionary relationship between the target and the template and by the quality of the sequence alignment. Both these factors can be estimated by the degree of similarity between the two protein sequences and by the number of insertions and deletions. In principle, the relationship between the sequence identity and the structural divergence between homologous proteins can be used to evaluate, a priori, the expected quality of the model, provided that we assume that the alignment is the correct structural alignment.

Even in difficult cases where the similarity between the target and template is quite limited, the mechanism of evolution plays on our side. The regions of the family that are better conserved and therefore easier to align and model are likely to be those playing an important structural or functional role. Therefore, notwithstanding all the caveats and problems associated with the homology modeling technique, it still represents an essential tool in modern biology and the models that can be obtained are a gold mine of biological information.

7.4 MANUAL VERSUS AUTOMATIC MODELS

What we discussed in this chapter especially applies when automatic modeling servers (i.e., publicly available programs that perform the complete model-building procedure automatically) are used, which is probably the most frequent case. Most of these tools have a very user-friendly interface; some have been carefully tested and evaluated and their output includes a summary of the expected accuracy for the different parts of the model. In some cases, the results of the intermediate steps of the procedure are made available to the user. It is of paramount importance that only reliable servers—those for which the results of blind independent tests are available—be used. It is also important to look at the intermediate steps, paying particular attention to the sequence alignment that, as we mentioned, is the crucial aspect of any modeling project.

7.5 PRACTICAL NOTES

Each of the steps of the homology modeling procedure introduces errors because of the underlying approximations that we have discussed at length so far. However, some technical details are not at all marginal, although they might seem trivial, and cause errors and frustration.

For example, we mentioned that it is very useful, whenever possible, to look at the structural superposition of proteins of the same evolutionary family. However, the problem of the optimal superposition between two structures is a complex one and involves the choice of some parameters that do affect the final result. For example, is it better to take into account a structural alignment where 50% of the structure is superimposed within a 1 Å r.m.s.d. or one that includes 80% of the structures, but results in a final r.m.s.d. of 1.8 Å? (Note that the superposition in the two cases is, in general, different.) There is no simple answer to this question. It depends upon what we want to use the superposition for. If we are trying to identify the common structural core, it is probably better to start from a superposition that includes most of the structures, analyze it, and iteratively remove parts of the two proteins that seem to deviate substantially from each other.

Also, the multiple sequence alignment methods are many and equally abundant is the set of tools to inspect and manually modify the alignment. A useful suggestion is to use different colors for each amino acid or type of amino acid because this allows misaligned regions to be easily identified.

When database search techniques are employed to model loops with insertions or deletions, a few parameters can be selected by the user. The alignment will indicate a number of amino acids to be inserted or deleted, but it is always advisable to search for regions including at least one more residue on each side of the insertion or deletion. For example, in a case such as the following, the best solution is to search in the database for regions with similar stems separated by two amino acids, to take into account what we discussed about the difference between a sequence and a structural alignment:

```
template  ADFRLADGTRCFRGT

target    LEFTLVD-SKCWKAS

          |stem|    |stem|
```

The size of the stems should also be decided on a case-by-case basis. For example, if the two stems have a helical conformation, it is preferable to use at least four or five residues to make sure that the selected local template also has a helical conformation; for β-strands, a shorter region can be sufficient. In all cases, it is always better to try more than one selection and verify how much the proposed solutions vary when the stem size is modified. Whatever the choice, though, a number of possible alternatives are possible and it is not clear how to select the most likely one. Again, we can try to give some rule of thumb, which can only be approximate.

The selected region should not collide with the main chains of the conserved regions of the model. If possible, the selected template loop should have the same pattern of glycines and prolines as that of the one to be modeled. If everything else is equal, the loop for which the structural similarity between the template and target stems is higher should be selected.

At this point, the modeler will meet another difficulty, which might seem minor (but is certainly not considered such by anybody who ever built a model)—that is, managing the protein databank files and, especially, their residue numbering. We have already briefly described the protein structure database entries. They look something like the following:

ATOM	49	N	ILE	Z	16	33.126	59.594	12.174	0.00	0.00	1	3TPI 155
ATOM	50	CA	ILE	Z	16	31.81	59.24	12.722	0.00	0.00	1	3TPI 156
ATOM	51	C	ILE	Z	16	30.843	60.42	12.748	0.00	0.00	1	3TPI 157
ATOM	52	O	ILE	Z	16	31.053	61.388	13.523	0.00	0.00	1	3TPI 158
ATOM	53	CB	ILE	Z	16	31.965	58.688	14.158	0.00	0.00	1	3TPI 159
ATOM	54	CG1	ILE	Z	16	32.941	57.499	14.229	0.00	0.00	1	3TPI 160
ATOM	55	CG2	ILE	Z	16	30.612	58.32	14.798	0.00	0.00	1	3TPI 161
ATOM	56	CD1	ILE	Z	16	32.349	56.23	13.587	0.00	0.00	1	3TPI 162
ATOM	57	N	VAL	Z	17	29.731	60.251	12.065	0.00	0.00	1	3TPI 163
ATOM	58	CA	VAL	Z	17	28.581	61.149	12.183	0.00	0.00	1	3TPI 164
ATOM	59	C	VAL	Z	17	27.533	60.535	13.105	0.00	0.00	1	3TPI 165
ATOM	60	O	VAL	Z	17	27.577	59.301	13.343	0.00	0.00	1	3TPI 166
ATOM	61	CB	VAL	Z	17	27.971	61.439	10.798	0.00	0.00	1	3TPI 167
ATOM	62	CG1	VAL	Z	17	28.965	62.183	9.886	0.00	0.00	1	3TPI 168
ATOM	63	CG2	VAL	Z	17	27.427	60.167	10.12	0.00	0.00	1	3TPI 169
....												
ATOM	1239	N	ALA	Z	183	20.936	65.758	26.824	1.00	19.04		3TPI1345
ATOM	1240	CA	ALA	Z	183	21.297	64.486	26.279	1.00	19.04		3TPI1346
ATOM	1241	C	ALA	Z	183	22.26	64.676	25.102	1.00	19.04		3TPI1347
ATOM	1242	O	ALA	Z	183	22.216	65.695	24.403	1.00	19.04		3TPI1348
ATOM	1243	CB	ALA	Z	183	20.071	63.625	25.866	1.00	19.04		3TPI1349
ATOM	1244	N	GLY	Z	184A	23.147	63.766	24.97	1.00	24.01		3TPI1350
ATOM	1245	CA	GLY	Z	184A	24.18	63.798	23.932	1.00	24.01		3TPI1351
ATOM	1246	C	GLY	Z	184A	25.595	63.678	24.5	1.00	24.01		3TPI1352
ATOM	1247	O	GLY	Z	184A	25.854	62.902	25.456	1.00	24.01		3TPI1353
	Atom number	Atom name	Res. name	chain	Res. number	X	Y	Z	Occ.	B-fact	Note	

Let us stress that the "residue numbers," notwithstanding their name, are character strings and not numbers; therefore, they are not necessarily consecutive and do not necessarily start from 1. They can contain letters (184A) because sometimes the numbering comes from a common numbering scheme of the evolutionary family of the protein. For example, the case shown here is that of the serine proteases where usually (but not always) the catalytic serine has the residue number 139 and the other two residues of the triad, the histidine and the aspartate, have the numbers 157 and 102, respectively.

The 3tpi entry of the databank contains the coordinates of the trypsin molecule together with those of a proteic inhibitor, so the enzyme is also identified by a letter representing the chain, in this case Z. In similar cases, even in the same family, the letter E is often used for the enzyme. Also, notice that residues 16 and 17 have occupancy 0 (i.e., no electron density is visible for these amino acids) and, indeed, the file contains a note:

```
AN OCCUPANCY OF 0.0 INDICATES THAT NO SIGNIFICANT
DENSITY WAS FOUND IN THE FINAL FOURIER MAP.
```

In the same file, there are sentences such as:

```
THERE IS NO SIGNIFICANT ELECTRON DENSITY IN THE FINAL

FOURIER MAP FOR THE N-TERMINUS OF THE ZYMOGEN FROM
VAL Z 10

THROUGH GLY Z 18 AND THIS DATA ENTRY CONTAINS NO

COORDINATES FOR VAL Z 10 THROUGH LYS Z 15.
```

This demonstrates the importance of reading the note records, which are unfortunately extremely difficult to parse automatically because they are in free text format.

Modeling programs tend to preserve the original template numbering scheme and that of the regions used to model loops. When database search methods are used for the latter, they do not verify the occupancy and do not record the original occupancy values of the protein atoms imported during the loop-building step.

Another possible problem can be encountered when the template or one of the proteins from where the coordinates of the loops need to be imported have been solved by NMR. As we mentioned, NMR does not produce a single structure, but a set of structures compatible with the experimental data. The databank file will therefore contain more than one structure or an "average" structure. Many programs solve the problem by simply discarding NMR structures in model-building procedures and it is difficult to criticize this choice.

7.6 SUMMING UP...

Consider the sequence of the following target protein:

```
DFTARTEDRDASRATTKSGSGTSHKLIPLPLFDERSEAWQRTAR...,
```

We search in the database of protein sequences for proteins of known structure that share a statistically significant sequence similarity with it:

```
Sequences producing High-scoring Segment Pairs:
Score
PDB|1QQR|PROTEIN XX          07-JUN-9...    43

...

PDB|1LUL|PROTEIN YY          30-JUN-9...    32
```

We obtain the respective sequence alignments:

```
>PDB| PROTEIN XX

Score = 43 Identities = 9/34 (26)

target  :    1 DFTARTEDRDASRATTKSGSGTSHKLIPLPLFDER 34
                 +FT R ++R+ +    K  SG + ++    L  E+
template:   90 EFTYRVKNREQAYRITKS-SGTSEEINNTDLISEK
123

>PDB| PROTEIN YY

Score = 26 (14.2 bits) Identities = 6/26 (23

target  :    3 TARTEDRDASRATTKSGSGTSHKLIPL 28
                 T+ T +  A    ++ +  G +   L+P+
template:   67 TSFTANIFAPNKSSSA-DGIAFALVPV 92
```

Let us now see how we build a model for a fragment of the target protein, given the alignment:

```
template   ...T K S - S G T S...

target     ...T S S G A G T A...
```

Let us recall that this is not the alignment proposed by database search programs such as FASTA or BLAST, but it has been optimized, possibly building a multiple-sequence alignment of as many members of the protein family as possible.

Let us now inspect the fragment in the context of the template structure (Figure 7.5) and copy the coordinates of its main chain atoms. Next, since we have to insert

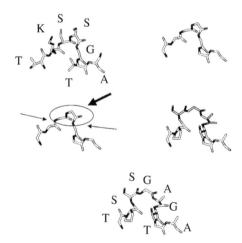

FIGURE 7.5 Scheme of a comparative modeling experiment.

an amino acid in the position indicated by the thicker arrow, let us search the database of proteins of known structure for loops that can be used as local templates. As we mentioned, although we only need to insert one amino acid, we will search for three amino acid regions whose stems will superimpose well with the stems of our model. The thinner arrows indicate their limits. In other words we will replace the amino acid included in the circle with three amino acids (Figure 7.5). Among the proposed solutions, we will select one; next, we will replace the side chains and optimize the model.

REFERENCES

Relationship between Sequence and Structural Similarity in Proteins

Chothia, C. and A. M. Lesk. 1986. The relation between the divergence of sequence and structure in proteins. *EMBO Journal* 5(4):823–826.

Crippen, G. M. 1977. Correlation of sequence and tertiary structure in globular proteins. *Biopolymers* 16(10):2189–2201.

Hilbert, M. et al. 1993. Structural relationships of homologous proteins as a fundamental principle in homology modeling. *Proteins* 17(2):138–151.

Definition and Analysis of Loop Regions

Ananthanarayanan, V. S. et al. 1984. Proline-containing beta-turns in peptides and proteins: Analysis of structural data on globular proteins. *Archives of Biochemistry and Biophysics* 232(2):482–495.

Chou, P. Y. and G. D. Fasman. 1977. Beta-turns in proteins. *Journal of Molecular Biology* 115(2):135–175.

Hollosi, M. et al. 1985. Studies on proline-containing tetrapeptide models of beta-turns. *Biopolymers* 24(1):211–242.

Milner–White, E. J. 1990. Situations of gamma-turns in proteins. Their relation to alpha-helices, beta-sheets and ligand binding sites. *Journal of Molecular Biology* 216(2):386–397.

Milner–White, E. J. and R. Poet. 1986. Four classes of beta-hairpins in proteins. *Biochemistry Journal* 240(1):289–292.

Morea, V. et al. 1998. Conformations of the third hypervariable region in the VH domain of immunoglobulins. *Journal of Molecular Biology* 275(2): 269–294.

Sibanda, B. L. and J. M. Thornton. 1985. Beta-hairpin families in globular proteins. *Nature* 316(6024):170–174.

Tramontano, A. and A. M. Lesk. 1992. Common features of the conformations of antigen-binding loops in immunoglobulins and application to modeling loop conformations. *Proteins* 13(3):231–245.

Tramontano, A. et al. 1990. Framework residue 71 is a major determinant of the position and conformation of the second hypervariable region in the VH domains of immunoglobulins. *Journal of Molecular Biology* 215(1): 175–182.

Wilmot, C. M. and J. M. Thornton. 1988. Analysis and prediction of the different types of beta-turn in proteins. *Journal of Molecular Biology* 203(1):221–232.

Wilmot, C. M. and J. M. Thornton. 1990. Beta-turns and their distortions: A proposed new nomenclature. *Protein Engineering* 3(6):479–493.

Sequence Alignment and Database Search Methods

Altschul, D. 1989. Gap costs for multiple sequence alignment. *Journal of Theoretical Biology* 138(3):297–309.

Gotoh, O. 1996. Significant improvement in accuracy of multiple protein sequence alignments by iterative refinement as assessed by reference to structural alignments. *Journal of Molecular Biology* 264(4):823–838.

Gracy, J. et al. 1993. Improved alignment of weakly homologous protein sequences using structural information. *Protein Engineering* 6(8):821–829.

Henneke, C. M. 1989. A multiple sequence alignment algorithm for homologous proteins using secondary structure information and optionally keying alignments to functionally important sites. *Computer Applied Bioscience* 5(2):141–150.

Kanaoka, Y. et al. 1989. Alignment of protein sequences using the hydrophobic core scores. *Protein Engineering* 2(5):347–351.

Koch, I. and T. Lengauer. 1997. Detection of distant structural similarities in a set of proteins using a fast graph-based method. *Proceedings of the International Conference on Intelligent Systems for Molecular Biology* 5:167–178.

Lesk, E. M. et al. 1986. Alignment of the amino acid sequences of distantly related proteins using variable gap penalties. *Protein Engineering* 1(1): 77–78.

Murzin, A. G. and A. Bateman. 1997. Distant homology recognition using structural classification of proteins. *Proteins* Suppl 1:105–112.

Notredame, C., D. G. Higgins, and J. Heringa. 2000. T-Coffee: A novel method for fast and accurate multiple sequence alignment *Journal of Molecular Biology* 302(1):205–217

Orengo, C. et al. 1992. Fast structure alignment for protein databank searching. *Proteins* 14(2):139–167.

Sander, C. and R. Schneider. 1991. Database of homology-derived protein structures and the structural meaning of sequence alignment. *Proteins: Structure, Function and Genetics* 9(1):56–68.

Subbiah, S. and S. C. Harrison. 1989. A method for multiple sequence alignment with gaps. *Journal of Molecular Biology* 209(4):539–548.

Thompson, J. D. et al. 1994. CLUSTAL W: improving the sensitivity of progressive multiple sequence alignment through sequence weighting, position-specific gap penalties and weight matrix choice. *Nucleic Acids Research* 22(22):4673–4680.

Side Chain Modeling

Abagyan, R. et al. 1997. Homology modeling with internal coordinate mechanics: Deformation zone mapping and improvements of models via conformational search. *Proteins* Suppl 1:29–37.

Cardozo, T. et al. 1995. Homology modeling by the ICM method. *Proteins* 23(3):403–414.

Chinea, G. et al. 1995. The use of position-specific rotamers in model building by homology. *Proteins* 23(3):415–421.

Dunbrack, R. L. and M. Karplus. 1993. Backbone-dependent rotamer library for proteins. Application to side-chain prediction. *Journal of Molecular Biology* 230(2):543–574.

Keller, D. A. et al. 1995. Finding the global minimum: A fuzzy end elimination implementation. *Protein Engineering* 8(9):893–904.

Laughton, C. A. 1994. A study of simulated annealing protocols for use with molecular dynamics in protein structure prediction. *Protein Engineering* 7(2):235–241.

Ogata, K. and H. Umeyama. 1997. Prediction of protein side-chain conformations by principal component analysis for fixed main-chain atoms. *Protein Engineering* 10(4):353–359.

Modeling of Insertions and Deletions

Bruccoleri, R. E. et al. 1988. Structure of antibody hypervariable loops reproduced by a conformational search algorithm [published erratum appears in *Nature* 1988; 336(6196):266] *Nature*, 335(6190):564–568.

Cardozo, T. et al. 1995. Homology modeling by the ICM method. *Proteins* 23(3):403–414.

Gerstein, M. and C. Chothia. 1991. Analysis of protein loop closure. Two types of hinges produce one motion in lactate dehydrogenase. *Journal of Molecular Biology* 220(1):133–149.

Jones, T. A. and S. Thirup. 1986. Using known substructures in protein model building and crystallography. *EMBO Journal*, 5(4):819–822.

Martin, A. C. et al. 1989. Modeling antibody hypervariable loops: A combined algorithm. *Proceedings of the National Academy of Science USA* 86(23):9268–9272.

Morea, V. et al. 1997. Antibody structure, prediction and redesign. *Biophysics Chemistry* 68(1–3):9–16.

Sanchez, R. and A. Sali. 1997. Advances in comparative protein-structure modelling. *Current Opinion in Structural Biology* 7(2):206–214.

Model Evaluation

Martin, A. G. et al. 1997. Assessment of comparative modeling in CASP2. *Proteins* Suppl 1:14–28.

Mosimann, S. et al. 1995. A critical assessment of comparative molecular modeling of tertiary structures of proteins. *Proteins* 23(3):301–317.

Moult, J. et al. 1995. A large-scale experiment to assess protein structure prediction methods. *Proteins* 23(3):ii–v.

Moult, J. et al. 1997. Critical assessment of methods of protein structure prediction (CASP): Round II. *Proteins* (Suppl 1):2–6.

Sali, A. et al. 1995. Evaluation of comparative protein modeling by MOD-ELLER. *Proteins* 23(3):318–326.

Samudrala, R. and J. Moult. 1997. Handling context-sensitivity in protein structures using graph theory: Bona fide prediction. *Proteins* Suppl 1:43–49.

Sanchez, R. and A. Sali. 1997. Evaluation of comparative protein structure modeling by MODELLER-3. *Proteins* Suppl 1:50–58.

Tramontano, A. and A. M. Lesk. 1992. Common features of the conformations of antigen-binding loops in immunoglobulins and application to modeling loop conformations. *Proteins* 13(3):231–245.

Venclovas, C. et al. 1997. Criteria for evaluating protein structures derived from comparative modeling. *Proteins* (Suppl 1):7–13.

Protein Structure Database

Bernstein, F. C. et al. 1977. The Protein Data Bank. A computer-based archival file for macromolecular structures. *European Journal of Biochemistry* 80(2):319–324.

PROBLEMS

1. Describe in "plain English" the meaning of the following line taken from a PDB entry.

22.8	ATOM	33	CG2	VAL	H	35	27.427	60.167	10.12	0.80	2XYZ 169

2. Given an alignment between two protein sequences, which regions are certainly not part of the conserved core?

3. Let us assume that we can obtain a sequence alignment that certainly reflects the correct evolutionary correspondence between amino acids of the two proteins. Discuss why this might still not represent a structurally correct alignment.

4. Why is the optimal structural superposition between two protein structures not unique?

5. Rewrite the following alignment in a more "structurally correct" way:

```
TRFTFTCVGTRERE

TKG-STCL--KERD
```

6. Discuss how can you use the information provided by the structure of the template for optimizing the alignment of its sequence with that of the target.

7. Discuss the advantage of finding more than one suitable template for a given target protein.

8. Most of what we discussed in this chapter relates to soluble proteins. Modeling of membrane proteins is more difficult. Can you find a couple of reasons for this to be the case?

9. Let us assume that we have produced a model of protein 1APH where all Cα of secondary structure elements are exactly at 1 Å distance from the corresponding atoms of the experimental structure and all other Cα are exactly at 2 Å. Compute the r.m.s.d. for the superposition of the Cα of the model and the experimental structure.

10. GDT-TS is a measure used in CASP for evaluating the quality of models and is defined as GDT_TS = (GDT_P1 + GDT_P2 + GDT_P4 + GDT_P8)/4, where GDT_Pn denotes percent of residues under distance cutoff less than or equal to n Å. Compute it for the previous example.

8 Fold Recognition Methods

GLOSSARY

Boltzmann's equation: equation linking the probability of observing the event x with energy $E(x)$ at a given temperature T. The higher the energy is, the lower the probability is: $P(x) = e^{-(E(x)/KT)}$. Ludwig Boltzmann was an Austrian physician who was born in 1844 in Vienna and died in Duino (Trieste, Italy) in 1906.

Fold library: set of structures that can be used as putative templates by fold recognition methods

Profile-based methods: fold recognition methods based on the comparison between the structural propensities of the amino acid sequence of the target protein and the properties of positions in proteins of known structure

Threading methods: fold recognition methods in which the target sequence is modeled using as templates each of the proteins in the fold library

8.1 BASIC PRINCIPLES

In previous chapters, we have seen that proteins belonging to the same evolutionary family share similar structures. However, the analysis of known protein structures has revealed that proteins apparently unrelated from an evolutionary point of view can also share a similar architecture.

Proteins sharing all or most of their secondary structure elements connected with the same topology are said to have the same fold. Lack of a significant sequence similarity among them can indicate that they are evolutionarily unrelated or that they have diverged so much from the common ancestor that their similarity fell below the detection limit of our sequence comparison methods. In the latter case, only their structural superposition can highlight some shared features pointing at a common evolutionary origin.

Independently of the reason or reasons, the fact remains that the number of folds so far explored by evolution seems to be finite. Therefore, it is highly likely that we will be able to see all possible folds without needing to solve the structure of all the proteins of the universe. This last point has important theoretical and practical implications; indeed, this chapter will discuss how it can be exploited for predicting the structure of proteins.

If a protein of interest does not share significant sequence similarity with any other protein of known structure, we can still try to see whether it has a fold that we have already observed in some other, apparently unrelated, protein. In other words, we can restate the problem of predicting a protein structure in terms of finding whether a given target sequence is compatible with one of the folds present in the database, regardless of its sequence similarity with any putative template. Such an approach is called "fold recognition" and includes two main techniques that we will briefly describe: profile based and threading.

8.2 PROFILE-BASED METHODS

As we discussed in previous chapters, each amino acid has different properties that determine the likelihood with which we can find it in different environments—for example, a hydrophobic or hydrophilic region, a given type of secondary structure, more or less exposed to the solvent, etc. The preference parameters for each amino acid can be estimated by counting its frequency of occurrence in the specific environment in known protein structures. For example, we can calculate, for each amino acid:

- how frequently it is found in a given secondary structure element (mostly alpha, mostly beta, or no preferences)
- how often it is found on the protein surface (high, low, intermediate)
- how often it is found in a hydrophobic environment (high, low)

As always, we will compute these figures, taking into account the expected random distribution.

There are 18 possible combinations of the propensities that we listed (3 for the secondary structure multiplied by 3 for solvent exposure multiplied by 2 for the hydrophobicity of the environment); therefore, each amino acid can be encoded with 1 of 18 characters and the sequence of a target protein recast into a different string where each character indicates the propensities of the corresponding amino acid for the selected set of environments (e.g., using the table shown in Figure 8.1). We can be more sophisticated and use prediction methods, for example, based on neural networks, for predicting these properties for the amino acids of our target sequence.

On the other hand, a protein structure can also be transformed into a linear string by replacing each of its positions, independently of which amino acid happens to be present, with a character indicating the secondary structure to which it belongs, how exposed it is, and what its environment is. In other words, we can ignore the specific sequence of amino acids of the protein and replace it with a character specifying the properties of each of its positions, as shown, for example, in Figure 8.2.

Note that, in this way, we are analyzing the protein in terms of its structure—not taking into account which amino acid occupies a given position, but rather which properties an amino acid should have in order to fit "reasonably well" in that position. For example, according to the code in Figure 8.2, the following string indicates a region of a protein structure where the first two positions are solvent exposed and therefore likely to be occupied by a hydrophilic amino acid and are in a β-strand and mostly surrounded by hydrophobic amino acids:

Mostly found in... / Frequency of presence on the surface	α	β	others
low	Most frequently found in hydrophobic environment (**a**) Most frequently found in hydrophylic environment (**d**)	Most frequently found in hydrophobic environment (**b**) Most frequently found in hydrophylic environment (**e**)	Most frequently found in hydrophobic environment (**c**) Most frequently found in hydrophylic environment (**f**)
high	Most frequently found in hydrophobic environment (**g**) Most frequently found in hydrophylic environment (**j**)	Most frequently found in hydrophobic environment (**h**) Most frequently found in hydrophylic environment (**k**)	Most frequently found in hydrophobic environment (**i**) Most frequently found in hydrophylic environment (**l**)
intermediate	Most frequently found in hydrophobic environment (**m**) Most frequently found in hydrophylic environment (**p**)	Most frequently found in hydrophobic environment (**n**) Most frequently found in hydrophylic environment (**q**)	Most frequently found in hydrophobic environment (**o**) Most frequently found in hydrophylic environment (**r**)

FIGURE 8.1 Possible encoding of amino acid propensities.

kkmdfaghjffacc

The third position is partially exposed to the solvent, in an α-helix and again surrounded mostly hydrophobic amino acids. Independently of the sequence that is actually observed in the protein of known structure, amino acids that have the properties corresponding to the substring kkm according to the encoding of Figure 8.1 can fit well in the substructure. By repeating this operation for each protein of known structure, we can transform the protein structure database into a "string" database.

The next step should now be obvious. We can use database search methods for detecting a similarity between the string representing the propensities of the amino acids of the target sequence and any of the string representing the specific environment of each position in known protein structures. A match with a known protein structure indicates that the properties of the amino acids of our target protein would be in a favorable environment if we used the known protein as a template.

Accessibility \ Secondary structure	α	β	others
<40 Å²	Hydrophobic (a) Hydrophylic (d)	Hydrophobic (b) Hydrophylic (e)	Hydrophobic (c) Hydrophylic (f)
>100 Å²	Hydrophobic (g) Hydrophylic (j)	Hydrophobic (h) Hydrophylic (k)	Hydrophobic (i) Hydrophylic (l)
intermediate	Hydrophobic (m) Hydrophylic (p)	Hydrophobic (n) Hydrophylic (q)	Hydrophobic (o) Hydrophylic (r)

FIGURE 8.2 Possible coding of amino acid properties within a protein structure.

The sensitivity of the method can be increased using multiple alignments of protein sequences of the same family.

8.3 THREADING METHODS

"Threading" is a term borrowed from the profession of tailoring, since *threading methods* essentially thread a sequence in the known structures. In other words, they build as many models as possible of a target protein, using all known structures as putative templates (even if they will only show a subset of them to the user). They try several alignments (including insertions and deletions) with each of the putative templates and evaluate which of the models, if any, is more likely to be correct.

The crucial step in this case is the evaluation of the quality of the fit between the target sequence and each of the putative template structures. It is computationally very expensive to carry out detailed energy evaluations for each of the many possible models. Furthermore, we expect that the target and potential templates share, at most, a similar fold, but are unlikely to be very similar in their details; therefore it would not be appropriate to use a very accurate energy evaluation method.

In fact, these methods use approximate energetic parameters describing the interactions of each amino acid rather than of each atom of the protein. The method used to derive such parameters is completely empirical. It can be shown that the frequency at which two residues are found within a certain distance in all known structures is correlated to the energy of their interaction. The relationship resembles *Boltzmann's equation*, which states that the probability of observing a given event depends on its energy:

$$P(x) = e^{(E(x)/KT)}$$

In other words, the higher the energy of an event (with respect to KT) is, the lower is its probability. This equation can be inverted and used to compute the energy of an event that is observed with a probability $P(x)$:

$$E(x) = -KT \ln[P(x)]$$

In a similar fashion, we can derive the interacting energy between two amino acids from the frequency with which they are observed to be in contact in experimentally determined protein structures with respect to a background random distribution.

We can use these energy parameters for estimating the energy of a protein structure by summing up all the appropriate energies for the pair-wise interactions that occur in the protein. If we do not know the specific alignment (i.e., where each amino acid of the target protein is placed in the structure as in the case of fold recognition methods), we need to compute it (i.e., find the correspondence between the target amino acids and the structure positions that minimize the total pair-wise energy).

This problem is somewhat similar to that of finding the alignment between two amino acid sequences that minimizes their difference (or maximizes their similarity) discussed in Chapter 3. Here, however, matters are more complex. Recall that, in the sequence alignment problem, the score that we use to fill the cell of the alignment matrix only depends upon the amino acid in the row and that in the column of the matrix. Here, the score also depends upon which amino acids occupy the nearby positions, since they will contribute to the pair-wise interaction energy associated with the amino acid.

The problem can be solved rigorously with a technique called double dynamic programming, which however is very computing intensive. More commonly, threading methods use the so-called "frozen approximation."

Let us assume that we want to compute the interaction energy associated to a valine in position 2 of a protein structure. We should compute the energy of its interactions with all other amino acids of the model in every possible position—that is, considering every possible target–template alignment. The rationale of the frozen approximation is that, whatever the final alignment is, it is likely that the amino acids of the target that will be placed in each position will have properties similar to those originally present in the template. Therefore, the interactions that they have with our valine will not be very different from those established by the amino acids of the template. Consequently, we can use, in first approximation, the amino acids of the template for computing the pair-wise interaction energies of the valine.

The frozen approximation is illustrated by the example in Figure 8.3. The interaction energy of the valine will be computed by adding the values of its pair-wise interaction parameters with the amino acids of the template, rather than with the amino acids of the target in each and every possible positioning of the latter in the template structure.

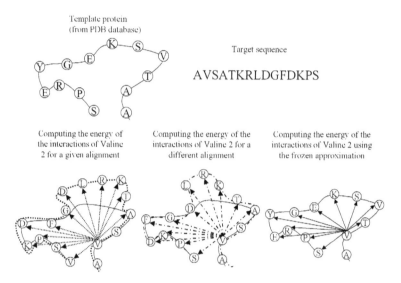

FIGURE 8.3 Simplified scheme of the frozen approximation.

8.4 THE FOLD LIBRARY

In profile-based and threading methods, we need to select the structures that we will use as putative templates—our *fold library*. Should we try all the known protein structures or only a subset of them—for example, selecting one structure for each evolutionary family? Should we use the complete structures or only the main secondary elements, neglecting the very variable loop regions? Should we construct a set of "idealized" folds or use the real ones?

There are methods employing each of these criteria and it seems, so far, that using a fold library as complete as possible is worth the effort. Nowadays, computers are sufficiently fast and inexpensive to make it easy to explore all the possible choices.

8.5 HOW WELL DO THESE METHODS WORK?

One somehow unexpected result of the first CASP experiments is the large variability in the performance of fold recognition methods.

In CASP, some fully automated homology modeling servers performed better than the average of the human predictors and very high quality predictions, together with some totally improbable ones, have been recorded for the same target. This is also the case for fold recognition methods. Some groups produce high-quality models; some fail even on easy targets. Sometimes, fold recognition models prove to be more similar to the experimental structure than any of the structures present in the protein structure database, suggesting that functional/structural intuition during model building is extremely important to optimize the quality of the model.

In difficult cases, it has been observed that, even if more than one group correctly predicts a given target, the same group rarely managed to predict all the targets correctly. This suggested that a possible way to improve the efficiency of the predictions is to combine different methods.

It should be considered, however, that the increased sensitivity of methods for sequence similarity searches, by now able to detect even very distant evolutionary relationships, has made the boundary between comparative modeling and fold recognition rather blurred. Indeed, starting from CASP8, targets sharing structural similarity with a known structure will be considered as a single set of targets, regardless of whether a clear sequence relationship is detectable or not.

REFERENCES

Historical Contributions

One of the first papers on fold recognition was that of Bowie, J. U., R. Luthy, and D. Eisenberg. 1991. A method to identify protein sequences that fold into a known three-dimensional structure. *Science* 253:164–170.

The papers stating that the number of protein folds is finite were produced by:

Chothia, C. 1992. Proteins. One thousand families for the molecular biologist. *Nature* 357:543–544.

Orengo, C. A., T. P. Flores, W. R. Taylor, and J. M. Thornton. 1993. Identification and classification of protein fold families. *Protein Engineering* 6:485–500.

Suggestions for Further Reading

Fold recognition methods are evaluated in the CASP experiments; hence, a more detailed description can be found in the special editions of *Proteins: Structure, Function and Genetics*.

The results of a workshop demonstrating the usefulness of combining different methods are detailed in:

Hubbard, T., A. Lesk, and A. Tramontano. 1996. Gathering them in to the fold. *Nature Structural Biology* 3:313.

Hubbard, T., J. Park, et al. 1996. Protein structure prediction: Playing the fold. *Trends in Biochemical Science* 21:279–81.

Hubbard, T., and A. Tramontano. 1996. Update on protein structure prediction: Results of the 1995 IRBM workshop. *Fold Description* 1:R55–63.

PROBLEMS

1. How would you compute the background distribution for the properties of amino acids mentioned in Figure 8.1?

2. Pair potentials are derived by computing the number of times two amino acids, say Ala and Val, are observed at a given distance from each other. Which other parameters should be taken into account to avoid a major bias in the results?

3. Search the Internet for fold recognition servers and catalogue them according to whether they are threading or profile based.

4. A user finds a clear match for the fold of a target protein with one of the methods described in this chapter. How should he or she go about establishing whether the two proteins are evolutionarily related?

5. How would you use the SCOP database for testing the sensitivity and specificity of a fold recognition method?

6. Approaches developed for fold recognition methods can also be used to assess the quality of a structural model obtained by different methods. Can you imagine how?

7. You are using a fold recognition method using as a query a protein of length M. You find a match with the structure of a protein of length N, with $N \gg M$. Is this more or less reliable than a match with the same score found in a protein with N almost equal to M? Why?

8. Your query protein does not seem to have any sequence similarity with any protein of known structure and therefore you recur to a fold recognition method. The result is a set of putative templates and their respective sequence alignment with the query. Which sequence similarity matrices would you use to score the alignments?

9. The diagrams in Figure 8.1P represent protein structures in two dimensions. Each circle is an α-helix and each square a β-strand. Which proteins have the same fold? What can you say about the relationship between the protein in Figure 8.1Pa and the protein in Figure 8.1Pd?

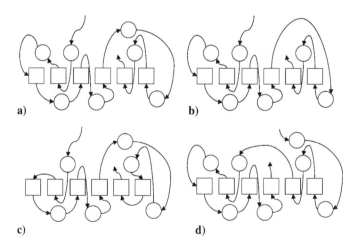

FIGURE 8.1P Schematic representation of protein architectures.

10. You are using a profile-based fold recognition method and your putative templates all have a secondary structure very similar to the one predicted by the Ph.D. program for your query protein. What should you verify about the fold recognition method to increase your confidence in the results?

9 New Fold Modeling

GLOSSARY

Force field or potential energy field: a set of parameters used to calculate the approximate energy of a protein

Genetic algorithms: optimization methods designed after natural evolution. The simulation starts using a large number of states, each of which can mutate or exchange parts with others. Elements with high "fitness" reproduce more effectively, so the next cycle of optimization starts with a population enriched in elements with higher fitness.

Levinthal's paradox: the average time for a process to find the state at minimum free energy is of the order of the time needed to explore all possible states. Even if we assume that each amino acid can only assume three states, a protein of 100 residues should explore 3^{100} conformations (something around 10^{48}). Even if it should only take 1 fs to explore a conformation, the protein would need 10^{25} years to find its native structure. Since this is longer than the life of the universe, a protein cannot reach its free energy state. This paradox is known as Levinthal's paradox, named after the scientist Cyrus Levinthal, who passed away in November 1990.

Molecular dynamics: algorithm that simulates the motion of the atoms of a protein given the forces acting on them

Monte Carlo methods: optimization methods based on a stochastic sampling of the solution space. The name was selected by Nicholas Metropolis during the Manhattan Project in the Second World War.

9.1 BASIC PRINCIPLES

The protein structure prediction methods described so far are heuristic; that is, they are based on the observation and analysis of proteins of known structure and cannot be used as such for proteins that do not share any structural similarity with proteins of known structure (i.e., for proteins that have a new, not yet observed, fold). Furthermore, even if these methods or novel ones that will undoubtedly appear in the field could provide us with accurate models of each of the proteins of the universe, we would still not be intellectually satisfied because proteins do not browse databases to achieve their native structure!

A protein folds into its native structure because the folded structure is the free energy minimum among the states that a protein can kinetically explore. It is energetically favored with respect to the unfolded state and to any other alternative conformation that can be explored by the protein. In principle, therefore, if we could generate all possible conformations of a protein and compute their respective free energy, the conformation with the lowest energy would correspond, with some exceptions, to its native state.

However, at least two problems make this strategy unfeasible. The first is that the stability of a protein is just a few kilocalories per mole (i.e., a couple of hydrogen bonds); therefore, even if we could generate all possible conformations, we should compute their energy with sufficient accuracy to appreciate energy differences of this order of magnitude. The second is that it is simply impossible to generate all the putative conformations of a protein. For a protein of 100 amino acids, even if we assume that each amino acid can only take three alternative conformations, there are 3^{100} different structures (10^{48}), and there is no way that we can generate and compute the energy of all of them.

This simple back-of-the-envelope calculation tells us that not even nature can afford to explore all the possible conformations of a protein. It would take 10^{25} years for a protein to do so, even if each conformation were sampled at the speed limit for physical processes ($1/10^{-15}$ s).

The theoretical treatment of the folding process that solves this apparent paradox (named after Levinthal) will not be described here. We will instead discuss the methods that we can use to evaluate the energy of a protein conformation and the strategies for generating a reasonable number of alternative conformations. We will also discuss another set of methods, named fragment-based methods. They take advantage of the observation that some structural features appear repeatedly in protein structures to reduce the size of the conformational space to be searched.

9.2 ESTIMATING THE ENERGY OF A PROTEIN CONFORMATION

A protein structure consists of a set of interacting atoms. Let us survey the atomic interactions that can be established among them and the approximations that we can use for estimating their energetic contribution to the overall energy of the protein.

A covalent interaction between two atoms (i.e., a chemical bond) keeps the atoms at an interatomic distance, which can vary around its equilibrium value r_0. A reasonable classical approximation for a covalent bond is to consider it as a spring between the two atoms. The spring will have an elastic constant, K_r, depending on the strength of the bond (e.g., a double bond will have a higher elastic constant than a single bond between the same types of atoms) and an equilibrium distance r_0. In this approximation, the energy associated with a covalent bond is given by:

$$E_{cov.bond} = K_r(r - r_0),$$

where, as we said, K_r and r_0 depend upon the types of atoms and their chemical bond.

Similarly, we can estimate the energy associated with the angle among three atoms using:

$$E_{angle} = K_\theta(\theta - \theta_0)^2.$$

The energy associated with the solid angle defined by four connected atoms can be approximated by:

$$E = \frac{1}{2} K_\varphi (1 + \cos n\varphi).$$

Another type of interaction between atoms is due to van der Waals forces. It is well known that this type of interaction is repulsive at short distances and attractive at longer distances. The van der Waals potential between atom i and atom j can be approximated by:

$$E = A_{ij} r_{ij}^{-12} - B_{ij} r_{ij}^{-6},$$

where r_{ij} is the distance between atom i and atom j. The first term represents the repulsive energy and the second the attractive one.

Charged atoms interact with each other. The electrostatic energy between atom i and atom j is given by the well-known Coulomb equation:

$$E_{electr.} = \frac{Q_i Q_j}{r_{ij}},$$

where Q_i and Q_j are the charges of the two atoms and r_{ij} their distance.

Electrostatic interactions decrease with distance less rapidly than other noncovalent interactions; hence, they tend to dominate the others. In the real world, soluble proteins are surrounded by molecules of polar solvent that shield the effect of atomic charges. This implies that we should include in our energetic calculations all the contributions of the water molecules and ions present in the solvent.

In macroscopic systems, the solvent effect is taken into account by writing:

$$E_{electr.} = \frac{Q_i Q_j}{\varepsilon r_{ij}},$$

where ε is the dielectric constant of the solvent.

This constant is a macroscopic variable that can be used to approximate the effect of molecules of the solvent in the hypothesis that the latter is infinite and isotropic. This is not the case for a protein solution, since the subject of our study, the protein, is of the same order of magnitude as the polar molecules. In early times when computer power was limited, protein energetic calculations used a dielectric constant rather than explicitly including all relevant water molecules and ions, but it should be kept in mind that this is quite a crude approximation. Nowadays, it is increasingly common to see simulations where solvent molecules are explicitly included in the system.

Hydrogen bonds are weak electrostatic interactions, so there should be no need to consider their contribution separately. However, the unavoidable errors introduced by treating a protein molecule as a classical object are sometimes insufficient to describe them. Therefore, many methods compute their energy separately:

$$E_{Hbond} = C_{ij}r_{ij}^{-12} - D_{ij}r_{ij}^{-10} .$$

Some methods can include other terms to take into account the energy associated with chirality, deviation from planarity, etc.

The set of terms

$$K_r, \ r_0, \ K_\theta, \theta_0, \ K_\phi, \ A_{ij}, \ B_{ij}, \ C_{ij}, \ D_{ij}, \ldots$$

for each pair, triplet, and quadruplet of atoms is called a *force field*. The parameters can be computed by statistically analyzing known molecular structures or using accurate *ab initio* quantum mechanical calculations on model systems. Given the force field and the structure of a protein, its total energy can be calculated by adding the terms described previously:

$$E = \sum_{bonds} K_r(r - r_0) + \sum_{angles} K_\theta(\vartheta - \vartheta_0)^2 + \sum_{\substack{solid \\ angles}} \frac{1}{2}K_\phi(1 + \cos n\varphi) +$$

$$\sum_i \sum_{j<i} A_{ij}r_{ij}^{-12} - B_{ij}r_{ij}^{-6} + \sum_i \sum_{j<i} C_{ij}r_{ij}^{-12} - D_{ij}r_{ij}^{-10}. \tag{9.1}$$

Clearly, the entire calculation includes several approximations. The error introduced by them is estimated to be of the order of 5–10%.

Keeping this in mind, we will now describe how we can try to minimize function (9.1) to look for the lowest energy (i.e., native) conformation of a protein.

9.3 ENERGY MINIMIZATION

The obvious way to find the lowest energy conformation among all the possible ones is to minimize function (9.1) with respect to the atom positions—that is, find which of the possible conformations corresponds to a lower value for the energy E.

There are several minimization methods. Intuitively, they work as if, starting from a given initial conformation, we were to vary the positions of the atoms and compute the energy associated to the new conformation. If the latter is lower than the starting energy, we accept the new conformation as a new starting one and repeat the operation; otherwise, we try a different move.

If the conformation with minimum energy is not very close to the starting one and they are separated by conformations with higher energy, we will not be able to reach it. This concept is expressed by saying that minimization procedures cannot

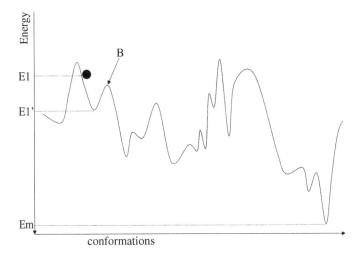

FIGURE 9.1 Hypothetical energy landscape of a protein.

guarantee finding the "global energy minimum" (i.e., the lowest energy minimum), but only "local minima" (i.e., conformations corresponding to an energy minimum that is not necessarily the lowest possible one.

Let us assume that we can "see" the whole conformational space of a protein and that we can plot the energy of each conformation as shown in Figure 9.1. If the starting conformation is the one indicated by "1" (with energy $E1$), the minimization procedure will find the conformation $1'$ (with energy $E1'$). The energy barrier B will prevent the algorithm from exploring the region where the conformation m, corresponding to the global minimum, is located.

From a structural point of view, we can visualize what happens by imagining, for example, a side chain on one side of another side chain in the starting conformation that should be on the other side in the global minimum structure. In order for the side chain to reach its minimum energy position, it should occupy positions such that its atoms would be very close to the atoms of the "interfering" side chain, and the repulsive contribution of the van der Waals potentials would be so high that the algorithm could only bring back the side chain to the original conformation after each move in the correct direction. Because of this, the global minimum would never be reached.

How can we solve the problem? We have two possible strategies. We can "push" our protein, by providing it with a kinetic energy that would allow the barrier to be overcome (*molecular dynamics*) or we can randomly explore alternative conformations without worrying whether a physically plausible way to go from one to the next exists (stochastic methods).

9.4 MOLECULAR DYNAMICS

Molecular dynamics is a technique that provides the energy needed for overcoming the potential barriers in the form of kinetic energy by assigning an initial velocity to each of the atoms of the protein and letting the protein move according to Newton's laws of motion. How do we assign the initial velocities? We can use the Maxwell–Boltzmann equation that states that the distribution of velocities in a gas is:

$$f(T,v) = 4 \times \Pi (m / 2\Pi kT)^{3/2} \exp(-\frac{mv^2}{kT})v^2 \,,$$

where
$f(T,v)$ is the fraction of particles with speed v at temperature T
k is the Boltzmann constant
m is the mass of the particles

Therefore, if we select a starting temperature for our system, we can assign velocities to each of the atoms in such a way that the distribution follows the Boltzmann law. The higher the temperature is, the higher will be the average kinetic energy and, consequently, the higher will be the barriers that the system will be able to overcome.

We can now treat our protein atoms as classical particles, each with a velocity, a starting position, and a force acting upon it. (The force will be the derivative of the energy that we can compute using Equation 9.2.) We can calculate the new position, velocity, and acceleration of each atom, after a time interval Δt, using Newton's law:

$$\vec{x} = \vec{x}_0 + \vec{v}\Delta t + \frac{1}{2}\vec{a}(\Delta t)^2$$

(9.2)

$$\vec{v} = \vec{v}_0 + \vec{a}\Delta t.$$

$$\vec{a} = \frac{\vec{F}}{m}$$

$$\vec{F} = \frac{\delta E}{\delta \vec{r}}.$$

It is well known that the preceding equations are valid for constant acceleration—that is, constant force. The force on an atom is not constant, since it changes as soon as the atom moves. However, if the time step Δt is very short, we can consider the force (and therefore the acceleration) constant. The procedure can be repeated to calculate position and velocity after $2\Delta t$, $3\Delta t$, etc. Clearly, the need of integrating the equation in very short time steps (of the order of the femtosecond) implies that a large amount of computation is needed to simulate the motion of the protein even for very short intervals of time.

The conformations generated during a molecular dynamics experiment are sampled at regular intervals and minimized. Their ensemble is subsequently analyzed to identify low-energy conformations and to analyze to dynamic behavior of the protein.

Presently, molecular dynamics is unable to simulate the folding process of proteins, which occurs on time scales of the order of milliseconds to seconds, but is nevertheless useful for exploring regions of the conformational space near the starting point. Perhaps more importantly, it is widely used to simplify the process of structure solution by techniques such as NMR and x-ray crystallography. In these cases, we can add a term to the energy in Equation 9.1 that takes into account the experimental data. For example, we can add a term that decreases the value of the energy when an atom is positioned within the experimental electron density obtained by an x-ray experiment or when two atoms are within the boundaries of a distance calculated in an NMR experiment.

The energy is then used for computing the force acting on each atom and molecular dynamics simulations can be used to move the atoms in positions more likely to satisfy the experimental constraints, while maintaining a reasonable stereo-chemistry, which is ensured by the terms present in Equation 9.1.

9.4.1 THE MONTE CARLO METHOD

Molecular dynamics follows a "time" course to search for possible conformations. Another option is to modify the positions of the atoms randomly and subsequently compute the energy of the newly obtained conformation. Since the conformational space of a protein, as we mentioned, is too large to be thoroughly explored, we need to limit our random search to "reasonable" conformations.

A class of methods known as *Monte Carlo methods* (named after the famous casino) is designed for this purpose. For example, the Metropolis algorithm explores the conformational space by assigning a low but finite probability of accepting moves (i.e., conformations) even if their energy is higher than that of the previous confor-mation. The rule is that the probability, P, of accepting a new conformation is given by:

$$P = \min\left(1, e^{\frac{-(E_{new}-E_{new-1})}{kT}}\right),$$

where k is the Boltzmann constant and T the temperature.

Let us see what the equation means. Assume that we randomly generate a conformation called *new* and calculate its energy. The new energy can be lower (case 1) or higher (case 2) than the energy of the starting conformation.

Case 1: $E_{new} \leq E_{new-1}$. The term $\dfrac{-(E_{new} - E_{new-1})}{kT} \geq 0$ (i.e., positive); there-

fore, $e^{\frac{-(E_{new}-E_{new-1})}{kT}} \geq 1$. Hence, $P = 1$; that is, the new conformation is always accepted.

Case 2: $E_{new} > E_{new-1}$. The term $\dfrac{-(E_{new} - E_{new-1})}{kT} < 0$ (i.e., negative); there-

fore, $e^{\frac{-(E_{new}-E_{new-1})}{kT}} < 1$.

$$P = e^{\frac{-(E_{new}-E_{new-1})}{kT}} .$$

In this second case, the higher the difference in energy is between the new conformation and the previous one, the lower is the probability of accepting the new conformation. For example, we can randomly change a dihedral angle and compute the value:

$$e^{\frac{-(E_{new}-E_{new-1})}{kT}} .$$

If this term is greater than one, the new dihedral angle is kept and a new move is performed starting from the new conformation. In all other cases, we generate a random number, N_{random}, between zero and one and the new conformation is accepted if and only if:

$$0 \le N_{random} \le e^{\frac{-(E_{new}-E_{new-1})}{kT}} .$$

This simple trick allows barriers of potential energy to be overcome since solutions with higher energy than that of the starting conformation can also be accepted. At the same time, the probability depends on the energy difference, so extremely unlikely conformations (i.e., conformations with very high energy) will not be accepted very frequently.

It is worth noting that the height of the barriers to be overcome depends upon a parameter that we called "temperature" and that we assign to the system. Generally, a high temperature is used for increasing the efficiency of the search of the conformational space. In some procedures (called simulated annealing), a high temperature is used at the beginning of the simulation, and the system is gradually cooled down during the simulation.

9.4.2 GENETIC ALGORITHMS

Genetic algorithms explore the conformational space of a protein by randomly generating a large number of alternative structures and selecting those that best fit our problem (i.e., that have a lower energy), analogously to what happens in Darwinian selection. Not only the general strategy, but also some of the details of these methods are modeled following the natural genetic mechanisms.

Let us describe the conformation of a protein as a string of bits reflecting, for example, the sequence of its ϕ and ψ angles. We generate a random initial ensemble of conformations and "operate" on the individuals of population by performing mutations and crossings-over. A mutation operation can, for example, exchange one bit with its complement (in our example, this corresponds to changing one dihedral angle). A crossing-over corresponds to the exchange of one part of one individual with the corresponding part of another one (in our example, we join the N-terminal fragment of one conformation with the C-terminal one of another and vice versa).

The fitness (the energy) of each "individual" (i.e., of the conformations encoded by the strings) is computed next and we select the individuals to be included in the next generation according to their fitness. Each individual will have a probability of reproducing related to its fitness; the lower the energy is, the higher is the probability of being selected for the next generation or the higher is the number of offsprings.

The procedure therefore generates a new population of individuals derived from the previous one, but more likely to have higher fitness, and the procedure is repeated. For example, the angles ϕ and ψ could be coded as:

ϕ (°)	ψ (°)	Number	Binary code
0–120	0–120	1	0001
0–120	120–240	2	0010
0–120	240–360	3	0011
120–240	0–120	4	0100
120–240	120–240	5	0101
120–240	240–360	6	0110
240–360	0–120	7	0111
240–360	120–240	8	1001
240–360	240–360	9	1010

A protein fragment could consequently be encoded by:

	Ala	Ser	Ser	Leu	Gly	Ala
ϕ (°)	118	300	270	250	40	260
ψ (°)	250	300	306	130	50	300
N	3	9	9	8	1	9
Bits	0011	1010	1010	0101	0001	1010

That is, it could be encoded by the string 0011.1010.1010.1001.0001.1010. A mutation could produce the string 0011.0010.1010.1001.0001.1010, implying that the Ser in position 2, previously in an a α_R conformation, has now assumed a different conformation. Could you see in which secondary structure the serine in the new conformation is? (A hint: In the Ramachandran plot, the angles are in the range of −180 to +180°; here, we used a different convention.)

A crossing-over between the generated individual and a new one encoded by **0011.0001.0001.0101.0111.0100** could produce the following two new individuals: 0011.0010.1010.**0101.0111.0100** and **0011.0001.0001**.1001.0001.1010, (where bold symbols refer to the second individual) and so on.

9.4.3 COMBINED METHODOLOGIES

As of today, unfortunately, none of the methods described in this chapter is able to find the global minimum conformation of a given protein consistently. Nevertheless, these algorithms and others with similar rationales can be used for predicting the structure of fragments of proteins or can be combined with other available information.

For example, we have discussed that, once we have predicted by homology the structure of the core of a protein, we can try to predict loop regions by searching possible local conformations at low energy that can connect the stems. Methods such as Monte Carlo and genetics algorithms can be useful to explore the conformational space of the fragments efficiently.

We do not need to limit this strategy to loops. A new set of methods has rather recently appeared and has met with remarkable success. In these methods, the protein sequence is split into nonoverlapping fragments. The often quoted Rosetta method uses fragments of three and nine amino acids. A number of fragments of proteins of known structure, identical or similar in sequence to each of the target fragments, are selected and their conformations are appropriately combined and optimized (for example, by Monte Carlo or genetic algorithms or other similar strategies). After a number of optimization rounds, the final list of models is scored according to an energy function, and the structure with the lowest energy is chosen.

In general, these methods use approximate residue-based pair-wise potential for evaluating the energy described in Chapter 8, often with the addition of a solvation term that tries to ensure that hydrophobic and hydrophilic residue are correctly partitioned between the surface and the interior of the protein. Secondary structure prediction methods are also used to bias the selection of the fragments of known structure so that, if a region of the protein is predicted to be helical, mostly helical fragments will be selected from the database.

These methods have been rather successful in CASP experiments and probably represent the most significant step forward in the structure prediction field in the last decade.

REFERENCES

Historical Contributions

The bibliography on the subject of molecular dynamics and energy minimization is really very rich and continuously growing. Some pioneer experiments are described in the following papers:

Levitt, M., C. Sander, and P. S. Stern. 1985. Protein normal-mode dynamics: Trypsin inhibitor, crambin, ribonuclease and lysozyme. *Journal of Molecular Biology* 181:423–447.

van Gunsteren, W. F., and M. Karplus. 1982. Protein dynamics in solution and in a crystalline environment: A molecular dynamics study. *Biochemistry* 21:2259–2274.

Suggestions for Further Reading

The following papers describe the use of molecular dynamics simulations in NMR and x-ray crystallography experiments:

Adams, P. D., N. S. Pannu, R. J. Read, and A. T. Brunger. 1999. Extending the limits of molecular replacement through combined simulated annealing and maximum-likelihood refinement. *Acta Crystallographica* D55:181–90.

Braun, W. 1987. Distance geometry and related methods for protein structure determination from NMR data. *Quarterly Reviews of Biophysics* 19:115–157.

Guntert, P., C. Mumenthaler, and K. Wuthrich. 1997. Torsion angle dynamics for NMR structure calculation with the new program DYANA. *Journal of Molecular Biology* 273:283–298.

Kraulis, P. J., and T. A. Jones. 1987. Determination of three-dimensional protein structures from nuclear magnetic resonance data using fragments of known structures. *Proteins: Structure, Function and Genetics* 2:188–201.

Stein, E. G., L. M. Rice, and A. T. Brunger. 1997. Torsion-angle molecular dynamics as a new efficient tool for NMR structure calculation. *Journal of Magnetic Resonance* 124:154–164.

van Schaik, R. C., H. J. Berendsen, A. E. Torda, and W. F. van Gunsteren. 1993. A structure refinement method based on molecular dynamics in four spatial dimensions. *Journal of Molecular Biology* 234:751–7623.

Fragment-based methods are described in:

Jones, D. T. 1997. Successful *ab initio* prediction of the tertiary structure of NK-lysin using multiple sequences and recognized supersecondary structural motifs. *Proteins* Suppl 1:185–191.

Simons, K. T., R. Bonneau, I. Ruczinski, and D. Baker. 1999. *Ab initio* protein structure prediction of CASP III targets using ROSETTA. *Proteins* Suppl 3:171–176.

PROBLEMS

1. Describe how you would design a profile-based fold recognition method taking advantage of structure and accessibility prediction methods.
2. How would you verify the likelihood of the frozen approximation assumption?
3. Discuss the meaning of the "temperature" parameter in the Metropolis Monte Carlo algorithm.
4. Design a different encoding for a protein structure to be used in a genetic algorithm.

5. Describe a fragment-based method where the optimization is based on a genetic algorithm.
6. Genetic algorithms can also be used for designing a sequence that fits a given structure (protein design). How would you do it?
7. Can you devise a way for exploiting the availability of a multiple-sequence alignment in a Monte Carlo optimization of a protein structure?
8. Why can Hooke's law approximate covalent bond energy?
9. The following table gives the approximate time scales of different motions in a protein. Compute how many molecular dynamics steps you would need to have statistical information about each process. (Assume that you want to explore, on average, 100 events in your simulations.)

Motion	Seconds
Bond stretching	10^{-14}
Angle bending	10^{-13}
Rotating CH3 groups	10^{-2}
Water tumbling	2×10^{-11}
Chemical reaction	10^{-6}

10. Why can you not use a dielectric constant of 80 (the value for water) for computing the Coulomb energy between two amino acid residues of a protein?

10 The "Omics" Universe

10.1 BASIC PRINCIPLES

Genomics is the science devoted to study of the genome. The genes of an organism are transcribed and, in general, translated into products, such as proteins that, in turn, assume a given three-dimensional structure and perform a function in the organism. We want to study them in a complete and "system"-wide fashion, understanding at which level, in which cells, and at which time the genes are transcribed (transcriptomics), of which proteins they direct the synthesis, where they are, how much there is of each of them, which post-translational modifications they undergo (proteomics), which structure they assume (structural genomics), the effect of a drug on each of them (pharmacogenomics), and so on.

We hear continuously of "omics" projects and bioinformatics is involved in each and every one of them not only for the obvious reason that they all produce a large amount of data but also, more importantly, because it is important to explore and take advantage of the inter-relationships between the data. In this chapter, we will shortly describe the most interesting "omics" projects and, even more briefly, we will touch upon the problem of how bioinformatics can theoretically contribute to each of them and what it can realistically achieve in the near future. (The two aspects, unfortunately, do not necessarily coincide.)

10.2 TRANSCRIPTOMICS

Almost every cell of an individual contains the same genetic information, but not all genes are active or expressed at the same level in all cells. In the past, the study of the activity of genes in different cells and at different times was limited to a few genes at a time, but a technological revolution—the development of microarrays—has completely changed the picture of what can and cannot be done.

A microarray is a support (e.g., glass), where single-stranded molecules of DNA are attached in specific positions (spots). Each microarray can contain tens of thousands of spots and each of them represents a single gene or a fragment of a single gene. The microarray can be designed for containing sequences identifying all known genes of an organism or by extracting the mRNA from a sample and retrotranscribing it into cDNA. The RNA of a sample (e.g., a cell culture) is extracted and labeled with a fluorescent molecule emitting, say, in the red wavelength. Similarly, the RNA of a control sample is extracted and labeled with a fluorescent molecule emitting at a different wavelength, say green.

The RNA of the sample and of the control is hybridized to the spots of the microarray. After washing, to eliminate nonhybridized molecules, the microarray is irradiated with a laser. If a specific RNA molecule is more (less) abundant in the sample with respect to the control, the corresponding spot will be red (green). If the RNA is equally abundant in the two samples, the spot will appear yellow; if neither of the two samples contains RNA complementary to the sequence of a spot, the spot will stay black. This implies that one can estimate the relative expression level of each of the RNA molecules from the intensity and color of the fluorescence. In practice, because of the limited stability of RNA molecules, these are retrotranscribed into DNA before the experiment.

The possible applications are many. In basic science, the integration of data in different organisms and cell types with sequence data and functional and structural information can increase our understanding of the studied system. We can formulate hypotheses on the function or on the metabolic pathways of genes that have a similar expression pattern. In applied science, the expression values of specific genes can be used for early diagnosis of a disease, for identifying correlation between diseases, and also to test whether a pathologically related level of expression can be normalized upon treatment with a drug.

The analysis of the expression data of normal cell populations that are chemically or pharmacologically perturbed or mutated in specific genes can allow us to understand the nature and the inter-relationships in the mechanisms of genetic regulation. This corresponds to the identification of groups of genes that have similar responses to the same perturbations (for example, addition of antibiotics, heat shock, growth in the absence of some nutrient, etc.). Figure 10.1 shows a schematic view of the results of a toy transcriptomic experiment. Our aim is to identify which of the spots (mRNA) have similar variations in the intensity (expression level) in different experiments. The first step is the quantification of the observed intensities.

The efficiency of extraction, labeling, hybridization, and washing of each RNA, as well as the number of cells in the sample, all contribute to the background noise in the data and can produce systematic errors in the results. The first problem is therefore the normalization of the data. The reference values used are usually the total amount of RNA or the intensity of spots of nonregulated genes (the housekeeping genes).

In the first case, we are assuming that the total amount of RNA in a cell is more or less constant, which is unfortunately not the case. The amount of the total and

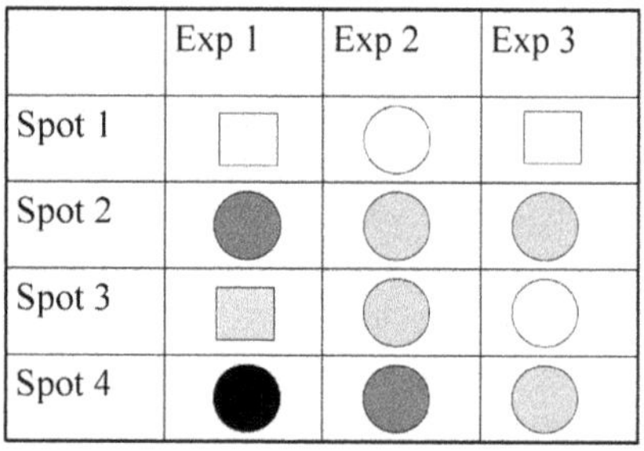

FIGURE 10.1 Results of a theoretical transcriptomics experiment. The circles (squares) tones correspond to RNAs more (less) expressed in the sample with respect to the control.

	Exp 1	Exp 2	Exp 3
Spot 1	-1	0	-1
Spot 2	3	1	1
Spot 3	-2	1	0
Spot 4	4	4	1

FIGURE 10.2 Possible quantification of the data shown in Figure 10.1.

Distance	Spot 1	Spot 2	Spot 3	Spot 4
Spot 1	0	21	3	45
Spot 2		0	26	10
Spot 3			0	46
Spot 4				0

FIGURE 10.3 Distance matrix derived from the data in Figure 10.2.

the ribosomal RNA can vary by up to two orders of magnitude. The expression level of housekeeping genes, such as actin, is less variable, although not really constant; therefore, this approach is appropriate unless the variations that we are interested in are minor (lower than a factor four or five).

The labeling efficiency can be taken into account by repeating the experiment, inverting the fluorescent label between the sample and the control. If normalization has been performed properly, Figure 10.1 can be transformed into a numerical matrix, where each cell represents the expression level of each mRNA molecule in a given experiment (Figure 10.2).

We can now define a "distance" between the different values—for example, by computing the sum of the squared difference between the values in the different experiments (Figure 10.3). The distance function can be selected to measure the actual difference in the expression levels of the genes with respect to a control (as shown in the figure) or it can be a correlation coefficient between the expression levels of two genes in different experiments, in order to capture similar behaviors of two genes even if their level of expression is different in absolute value.

The problem is very similar to the one we faced when we were grouping genes or proteins according to their sequence similarity. Here, as in that case, the distance values can be used to build a tree. In Chapter 3, we used it to cluster genes of proteins with similar sequences; here we use it to cluster genes whose expression level varies in a similar fashion in different experiments. The topology of the tree can help us cluster our clones as shown in Figure 10.4.

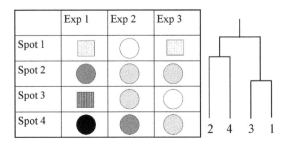

FIGURE 10.4 Clustering of the spots. The tree shown on the right can be computed from the data in Figure 10.3.

Once again, though, we need to ask which of the correlations that we observe are statistically significant with respect to what we expect by chance. We can use bootstrapping techniques, but in this case we have much more significant errors associated with the data, problems with their reproducibility, and, in general, lack a model for the observed phenomenon.

In identifying the relationships between genes, we can use other information at our disposal, integrating the expression data with the information stored in biological data banks. This last approach is the most attractive but also rather complex; clustering algorithms are not very easy to integrate with different data types.

In general, the methodology for data analysis in transcriptomics experiments is still being extensively studied. On the other hand, the transcriptional responses of many genes in different tissues of different organisms, in various conditions, and upon different treatments are accumulating very fast in public databases. We still do not know, in many cases, the specific phenomenon responsible for the observed pattern of expression and this limits the effectiveness of our analyses. However, the availability of the data is the first essential step for the development of methods that will allow us to investigate the complex network of gene regulation.

10.3 PROTEOMICS

Assuming that the amount of produced protein is proportional to the expression level of the encoding gene is a rather crude approximation. The control of the amount of protein present in a given cell at a given stage can depend upon factors other than its transcriptional level. For example, the thermodynamic stability, and therefore the half-life, of the corresponding mRNA molecule can play an important role in establishing the amount of produced protein. Furthermore, the transcription level of a gene does not provide any information about putative post-translational modifications of the product, which can play essential roles in regulating its activity.

The obvious solution is to inspect directly the protein content of a cell, rather than its mRNA content. A classical electrophoresis of the total protein content of a cell is unfortunately useless for this purpose because the number of proteins in a cell is such that it would be impossible to distinguish and identify each of the molecular entities only on the basis of its molecular weight. It is, however, possible to recur to the so-called bidimensional gels, where the proteins are separated first according to their molecular weight and subsequently according to their isolelectric point. This increases the probability that each spot on the gel corresponds to a single protein.

Analogous to the case of transcriptomic experiment, we can compare the relative abundance of a protein in a different experiment and try to obtain information about its function, but we need to be able to identify the protein. Once again, technology has given us an essential aid with mass spectrometry techniques, especially tandem mass spectrometry.

Mass spectrometers are analytical tools used for measuring the molecular weight of a sample. They are formed by three fundamental parts: the ionization source, the analyzer, and the detector.

The sample is introduced into the ionization source of the instrument where the molecules are ionized. These ions are extracted into the analyzer region of the mass spectrometer, where they are separated according to their mass (m)-to-charge (z) ratios (m/z). The separated ions are detected and this signal sent to a data system where the m/z ratios are stored together with their relative abundance. The sample can be inserted directly into the ionization source or can undergo some type of chromatography—for example, high-pressure liquid chromatography, gas chromatography, or capillary electrophoresis.

The ionization methods are many. Probably the most used in proteomics are electrospray ionization (ESI) and matrix-assisted laser desorption ionization (MALDI). In ESI, the sample is dissolved in a polar, volatile solvent and pumped through a narrow, stainless steel capillary. A high voltage is applied to the tip of the capillary, which is situated within the ionization source of the mass spectrometer; as a consequence of this strong electric field, the sample emerging from the tip is dispersed into an aerosol of highly charged droplets. Eventually, charged sample ions, free from solvent, are released from the droplets; some of them will pass through a small opening and make it to the analyzer of the mass spectrometer.

MALDI is based on the bombardment of sample molecules with a laser light to ionize the sample, which is premixed with a highly absorbing matrix. The matrix transforms the laser energy into excitation energy for the sample, which leads to sputtering of analyte and matrix ions from the surface of the mixture. The mass analyzer separates the ions formed in the ionization source according to their mass-to-charge (m/z) ratios.

Tandem (MS-MS) mass spectrometers are instruments that have more than one analyzer (usually two). The two analyzers are separated by a collision chamber where an inert gas (e.g., argon, xenon) collides with the sample ions and fragments them. The first analyzer is used to select user-specified sample ions arising from a particular component. Next, the ions are analyzed (i.e., separated according to their mass to charge ratios) by the second analyzer. All the fragment ions arise directly from the precursor ions specified in the experiment and thus produce a fingerprint of the compound under investigation. Tandem MS-MS mass spectrometers can be used for peptide and nucleotide sequencing and therefore to identify proteins in a proteomics experiment.

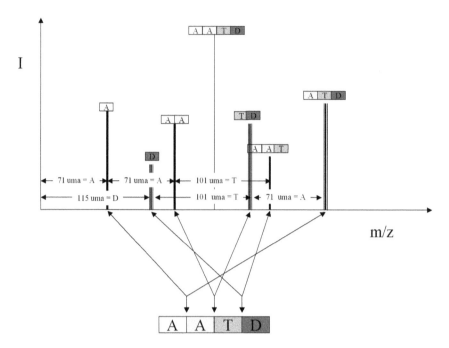

FIGURE 10.5 Example of the use of tandem mass spectrometry for obtaining the sequence of a protein fragment. In the example, we assumed that the initial fragment has a double charge. The peaks corresponding to the "b" series are indicated by single lines and those of the "y" series with double lines. The mass difference between two consecutive peaks of the same series allows the identification of the amino acid separating them. In a real case, we do not know which peaks belong to the "y" and "b" series.

Three bonds can break during fragmentation: the NH–CH, CH–CO, and CO–NH bonds. Each bond breakage gives rise to two species—one neutral and the other one charged—and only the charged species is monitored by the mass spectrometer. The charge can stay on either of the two fragments. Hence, there are six possible fragment ions for each amino acid residue and these are labeled "a," "b," and "c" ions when the charge is retained on the N-terminal fragment, and "x," "y," and "z" ions when the charge is retained on the C-terminal fragment. The most frequent cleavage occurs at the CO–NH bonds, giving rise to "b" and/or "y" ions. The distance between the peaks of the "y" and "b" series allow us to deduce the sequence of the fragment (Figure 10.5).

Once we isolate the portion of the gel corresponding to the protein of interest—literally cutting the portion of the gel—we can study the material and sequence the protein responsible for the spot of interest. Even if we are lucky and the amount of protein in the spot is sufficient and the spot only contains one species, there are other technical problems to be solved. As we can see from the figure, the sequence of the fragment can be obtained only if we solve a combinatorial problem, since we do not know a priori which peaks correspond to the "y" series and which to the "b" one.

Next, given the sequence of some of the protein's fragments, we need to identify it unambiguously. Some methods attempt to reconstruct the complete protein sequence using various enzymes to generate overlapping fragments of the protein; others search biological databases for proteins containing the identified fragments,

In the latter case, it is obviously essential to search a data bank as completely as possible and to be able to correctly evaluate the probability that the identity between the sequence produced by the mass spectrometer and that of the fragment of the database protein is statistically significant, However, the peptide sequences are short and can be present in different proteins; therefore, the significance of the score is difficult to evaluate. Fortunately, the experiment will provide us with the sequence of more than one fragment of the sample protein.

10.4 INTERACTOMICS

Proteins do not act alone, and therefore understanding their function requires the knowledge of their interaction partners. Experimental techniques such as affinity chromatography, coimmune precipitation, yeast two-hybrid systems, etc., can provide a map of the interaction for all (or most) proteins in a cell. These methods are very powerful and fast, but have limited specificity; that is, they often detect spurious interactions. This should not come as a surprise because protein interaction is not an "all or none" phenomenon. Proteins can interact with different affinity in different conditions and some proteins never interact simply because they are sequestered in different cellular compartments.

We mentioned computational methods for detecting protein–protein interactions in Chapter 6. Let us recall that they can be based on the identification of cases where two interacting proteins are separated in one species but are present as two domains of the same protein in the other or on comparative genome analysis to detect co-occurrence of proteins in different genomes. Similarity in the rate of evolution can also be used for predicting protein–protein interactions. If two protein families show the same rate of evolution (i.e., if the trees built on the basis of their sequences are similar in their topology and in the length of their branches), this might be due to coevolution brought about by the presence of an interaction.

All these methods have been applied to derive maps of putative protein interaction networks in organisms. These maps are not yet sufficiently reliable to be used other than as a guide for designing experiments. The limited reliability of the experimental data sets is reflected in the low accuracy of prediction methods, which need to use the experimental data for training and testing.

Another important aspect to keep in mind is that theoretically and experimentally derived interaction maps are static representations of complex and dynamic phenomena. In general, they are not able, at present, to distinguish between mutually exclusive interactions and, in many cases, between direct and indirect interactions (i.e., interactions between two proteins mediated by the presence of other proteins).

10.5 STRUCTURAL GENOMICS

The problem of assigning a molecular function to each and every protein cannot be solved without taking into account the three-dimensional structure of the protein. The astonishing progress of x-ray crystallography and NMR spectroscopy and the possibility, especially in crystallography, to automate most of the steps of the procedure opened the road to the possibility of solving in a high-throughput mode the structure of a large number of proteins.

The international community has therefore set up a number of large projects aimed at solving the structure of as many proteins as possible in the shortest possible time. Even so, it is unthinkable that we can experimentally solve the structure of all the proteins of the universe.

On the other hand, as we have repeatedly mentioned, the structure of a protein is conserved through evolution. Therefore, it is reasonable to rationally select the proteins to analyze among those belonging to evolutionary families for which we have no structural information. The structure of a protein of the family will provide us with a template for predicting by homology the structure of the other members of the family. This, by and large, is one of the objectives of structural genomics projects.

For this to be successful, two conditions must be met: the correct identification of the proteins belonging to the same evolutionary family (i.e., a correct selection of the sequence identity threshold above which two proteins can be predicted to be homologous) and good alignments that can be used to produce reasonably accurate homology models. The latter is a rather complex problem and there are indications that sequence identity between target and template might not be the best parameter for evaluating *a priori* the expected quality of a model.

Another nontrivial problem is that of multidomain proteins, which are very difficult to deal with. Even if two multidomain proteins have a high sequence similarity along their complete sequence, the relative orientation of the domains can vary substantially during evolution. Finally, let us not forget all the problems connected with obtaining a correct sequence alignment.

Other structural genomics projects aim at solving the structures of all the proteins of a given genome. This goes in the direction of collecting all the relevant information about one specific organism. The subsequent step would be the accurate simulation and understanding of all the cellular functions. This might seem a very ambitious goal, but, given the speed at which new challenges are met every day, even this dream could become reality within a relatively short time frame.

10.6 PHARMACOGENOMICS

The development of a drug is a far more complex task than just finding a molecule able to interfere with a given molecular or cellular function. Once a target (i.e., the molecule or the process with which we want to interfere) has been identified and once an assay for testing the activity of potential inhibitors has been set up—and even when a compound with the desired activity has been identified (lead compound)— we are only at the beginning of a long route toward the development of a drug.

The lead compound is only one of the many compounds that will be synthesized and tested in our assay during the drug discovery process. We will need to optimize several other parameters; the biological activity is only one of them. The final molecule also has to be nontoxic, amenable to chemical synthesis and large-scale production, have an appropriate half-life and metabolic destiny, not be sequestered by proteins other than its target, be absorbable, reach the cell type and compartment where the target resides, etc.

Once a molecule with all these properties has been isolated, preclinical studies can start. These are studies on animal models aimed at evaluating the safety of the compound. If all goes well up to this point, the molecule can be investigated in clinical trials, which are usually done in four phases:

- Phase I: The compound is tested on healthy volunteers. This permits verification that the metabolic and safety data obtained in animal models can be extended to humans.
- Phase II: The compound is tested on volunteering patients. This evaluates its efficacy and optimizes its dosage.
- Phase III: The compound is tested on a few thousand patients. This compares its efficacy with respect to other treatments, if available, or with placebos. If an effective advantage of the compound is established and there are no serious adverse effects, the drug can be introduced into the market.
- Phase IV: Even after the introduction of the drug into the market, its efficacy and safety are monitored. The appearance of unexpected adverse side effects can lead to counterindication for its usage or even to its removal from the market.

For a pharmaceutical company, a failure in Phase III or, even worse, in Phase IV is expensive. On the other hand, even if all the previous steps have been performed in an exemplary fashion, the risk of unexpected side effects exists for any drug. It is impossible to test all possible combinations with all other available drugs, in all possible physiological or pathological conditions, or the effect on patients with different genetic backgrounds.

A tool that would allow the effects of the drug on patients to be evaluated beforehand and to be correlated with genetic backgrounds would be of enormous economical value but, above all, could avoid problems to patients. It is intuitive that, at least in principle, we could use gene expression data to try to correlate the effects of a drug on a cell type and/or to correlate specific expression patterns with the toxicity of a compound.

This is an objective of extremely high relevance for applied science as it is testified to by the appearance of a plethora of new companies that aim at tackling this problem. A World Wide Web search for the keyword "pharmacogenomics" finds thousands of pages; the large majority correspond to Web sites of biotechnological companies.

10.7 BUT THIS IS NOT ALL…

The enormous interest in "omics" projects is not only due to the need for collecting data and derive databases, but also to the fact that the understanding of the functional properties of even a single molecule cannot be obtained out of its context and that, only with a synergic effort, can we hope to understand the complex network of biological interactions.

Keeping pace with the "omics" terminology is a formidable challenge. We have listed some more commonly used terms in Table 10.1. Unfortunately, the frequency with which new terms are invented is such that often the same term is used in different laboratories to indicate different things.

The existence of an "omic" word does not always imply that a project to analyze the data is already running, but we can be sure that none of these projects, ongoing or just foreseen, can be successful if it cannot take advantage of appropriate and effective bioinformatics tools. Therefore, let us go back to work!

TABLE 10.1
Definitions of Some "Omics" Terms

Term	Study of:
Biomics	Major regional community of plants and animals with similar life forms and environmental conditions
Cellomics	Set of all states that a cell could enter
Epigenomics	Genome-wide distribution of methylated and unmethylated nucleoside residues within the genome
Glycomics	Structure and function of oligosaccharides
Interactomics	Complete set of macromolecular interactions, physical and genetic
Immunomics	Molecules that are part of the immune system; immunodominant epitopes in an organism
Metabolomics	Unique chemical fingerprints of specific cellular processes (i.e., their small-molecule metabolite profiles)
Operomics	Proteins of unknown function; integrated genomic and proteomic profiling of cells and tissues
Phemomics	All possible phenotypes
Physiomics	Complete set of interactions that produce the physiology of an organism
Regulomics	Genome-wide regulatory network of the cell
Somatonomics	All somatic gene rearrangements
Chemogenomics	Gene family-based approach to drug discovery and target validation

Useful Web Sites

We list here only a few Web site URLs: those more commonly used and, especially, that are sufficiently stable. All other useful Web sites can be usually reached using links from the pages listed here.

National Library of Medicine of the United States contains several databases, integrated systems, and tools: http://www.ncbi.nlm.nih.gov

European Bioinformatics Institute (EBI) contains databases and tools: http://www.ebi.ac.uk

Sanger Centre is devoted to genomics and postgenomics research: http://www.sanger.ac.uk

Genome Data Base GDB: http://gdbwww.gdb.org

Japan's genome site: http://www.genome.ad.jp

A site for the analysis of protein sequences and structures: http://www.expasy.ch/

European Molecular Biology Laboratory (EMBL): http://www.embl.de

The official site of the protei structure database: http://www.rcsb.org/pdb/

Cambridge Crystallographic Data Centre (includes the structures of small organic and organometallic molecules): http://www.ccdc.cam.ac.uk/

The structural classification databases:
  http://scop.mrc-lmb.cam.ac.uk/scop
  http://www.biochem.ucl.ac.uk/bsm/cath
  http://www2.ebi.ac.uk/dali/fssp/fssp.html

International Society for Computational Biology: http://www.iscb.org

Center for Molecular and Biomolecular Informatics in The Netherlands:
  http://www.cmbi.kun.nl
  http://bioinfo.pl/LiveBench/

The official CASP site: http://predictioncenter.org

Hidden Markov models: http://hmmer.wustl.edu

Some free (or almost free) molecular graphics programs:
  http://www.cmbi.kun.nl/whatif/
  http://www.umass.edu/microbio/rasmol
  http://sgce.cbse.uab.edu/ribbons/
  http://www.expasy.org/spdbv/
  http://trantor.bioc.columbia.edu/grasp/
  http://www.biochem.ucl.ac.uk/bsm/ligplot/ligplot.html

Some commercial molecular graphics programs:
   http://www.tripos.com
   http://www.accelrys.com

Some docking programs:
   http://www.cmpharm.ucsf.edu/kuntz/dock.html
   http://cartan.gmd.de/flexx/

Molecular dynamics and minimization:
   http://igc.ethz.ch/gromos/
   http://www.gaussian.com
   http://www.gromacs.org/

Databases containing metabolic and regulatory networks can be reached starting
   from: http://www.ebi.ac.uk/research/pfmp/texts/biochemical_networks_web.
   html

Index

171